Cosmology

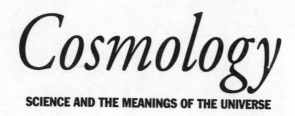

Cosmology

SCIENCE AND THE MEANINGS OF THE UNIVERSE

JOHN McLEISH

BLOOMSBURY

Also by John McLeish
Number

The Publisher would like to thank Eamonn Kerins for his help with the line
illustrations in this book.

First published 1993
by Bloomsbury Publishing Limited, 2 Soho Square, London W1V 5DE

Copyright © 1993 by John McLeish

A CiP record for this book is available from the British Library

The right of John McLeish to be identified as the author of this
work has been asserted by him in accordance with the Copyright,
Designs and Patents Act 1988

ISBN 0 7475 1145 4

Typeset by Hewer Text Composition Services, Edinburgh
Printed and bound by Clay Ltd, St Ives plc, Bungay, Suffolk

Contents

Prologue   ix
   Development of twentieth-century cosmology   x

PART I   THE SCIENTIFIC FOUNDATIONS

Chapter 1 The Origins of Cosmology
   Twentieth-century lift-off   3
   Cosmology and early societies   5
   Prehistoric science   7
   The Chinese philosophy of science   9
   Chinese cosmology   12

Chapter 2 Greek Cosmology and its Aftermath
   Platonism   17
   Aristotle's legacy   20
   Ptolemy's system   24
   Physical science in pre-modern Europe   26

Chapter 3 The Take-Over by Physics
   Cosmology as an exact science   29
   The science of dynamics: the inside track   30
   Information from far-away places   38
   Obtaining celestial data   39
   Measuring distance   44
   Measuring orbital velocity   47
   The chemical composition of the universe   48

Chapter 4 Continuity in Nature

Classification: the basic method of science 51

How old is the Earth? 51

Radioactive dating 53

The origin of the solar system 55

Sky objects – the raw materials of cosmology 57

Galaxies 58

Nebulas 61

Star clusters 62

Star names and the history of cosmology 63

Classification of star spectra 64

Recycling the ether – the Michelson–Morley
experiment 66

PART II   REDEFINING THE UNIVERSE

Chapter 5 Quantum Theory: A New Look at Energy

A new beginning 73

Ernst Mach, forerunner of the new physics 74

Opening chorus – the trouble with science 77

Determinism and prediction 77

On the nature of light 78

Duality in wave-particles 80

The Copenhagen manifesto 81

Indeterminacy: Bohr vs. Einstein 84

Heisenberg: the new quantum theory 86

Dirac's contribution 89

Some nonsensical interpretations 91

Philosophy and physicists 92

Chapter 6 Relativity: The Fast-Moving Universe

Einstein's relativity: phase 1 (1905–1911) 95

Relativity phase 2: the established principles 99

Relativity phase 3: gravity and the time–space
continuum 103

The geometry of space–time 104

Gravity waves 105

CONTENTS

Slipher, Hubble and the expanding universe   107
Gravity as a field   108

PART III   THE INVISIBLE UNIVERSE

Chapter 7 Smashing Atoms: Looking at the Smithereens
Building on the past   113
A tribute to Democritus   113
Science and the atom   114
The periodic table and atomic structure   117
The periodic system of elements   119
Elements and their isotopes   121
The quark – smallest particle of matter?   123
Targeting the nucleus   124
The electron microscope – looking into the nucleus?   128
The secret of the Sun: nuclear power   129
Splitting the atom – nuclear fission   129
Nuclear fusion   130
Where does nuclear energy come from?   131

Chapter 8 A Cosmic Interlude
Looking forward and back   133
The Hertzsprung–Russell chart   135
The rebirth of relativity   136
The four forces of nature   137
Curved space–time   140
The 'field' concept of gravity   141
Energy particles and waves: light and gravity   142
Dark matter and black holes   142
Creation – a 'Big Bang' or just a continuous hiss?   145

Chapter 9 Energy Sources in the Sky
Setting the scene: a personal memory   148
The invisible world of radio waves   149
The principle of the interferometer   152
Gamma rays and their sources in the universe   154

Photons – bearers of light and electric power 156
X-rays in the universe 158

PART IV HUMAN INTEREST

Chapter 10 The Search for Life in the Universe
Abandoning some cherished illusions 163
Defining life 164
Is there life on Mars? 167
The Red Planet – close encounters 171
Testing Mars for signs of life 172
Is there any extraterrestrial life at all? 173
Radio contact with outer space 176

Chapter 11 In Place of a Summary
Cosmology and speculation 179
Matter, fields and energy: a 10-point summary 181

Chapter 12 Science and Religion
The religious origins of Western science 185
The Hebrew myth of Creation – a personal view 187
God in theology 187
Arguments for God's existence: a critique 191
Can God exist in a scientific age? 196

Chapter 13 The New Cosmology: Future Prospects
String models – building a universe 198
Evidences of invisible matter 202

Prologue

From early times until fairly recently, the human species has been absorbed in problems of personal and social survival. Very quickly we developed a method for dealing with the unknowable. Confronted by a mystery, we fabricate a story as our first, normal reaction. We still do this when we have no ready magic for finding a solution. The myth dispels anxiety, and the story can be embellished later and articulated as a work of art. This is the origin of creation myths: they are an early art form. Myths were dominant cultural themes in literature and art before some of them became enshrined in religious dogma.

The Creation theme has many facets. Over time, methods are developed for clarifying, verifying and even forgetting unacceptable variants. Mythology flourished from the start as a defence structure. It also came to be accepted as a test of allegiance, having been converted into an ideology with its own standards of power, convenience, plausibility and – a long way down the road – truth. For a long time, mythology offered the only framework for a theory of the cosmos. For many people, this is still the only kind of cosmology that they know.

A second kind of cosmology arose out of the first. The scientific method was devised by the peoples of the ancient Middle East – Phoenicians, Babylonians, Egyptians, Arabic-speaking nations – and by the Chinese. As johnnies-come-lately, the Greeks claimed the whole art of thought as their invention. In cosmology, Galileo, above all others must be given credit for challenging this Greek claim to a monopoly of thought, which Roman Christianity had presented as its own and imposed on Europe. In the course of his struggle for a new rational basis for knowledge, Galileo invented cosmology as a science.

Galileo's magnificent work, *A Dialogue Concerning the Two Great World Systems, Ptolemaic and Copernican*, was published in 1632. In it he states the founding principle of cosmology: that there is no substantial difference between the so-called 'heavenly bodies', and Earthlike so-called 'sub-lunary' bodies such as the planets. Once declared and accepted, this principle chartered scientists to perfect their findings 'on the ground', in their laboratories, before using the results in their study of the heavens. This cosmology, and its development in the twentieth century, are the subjects of this book.

Development of twentieth-century cosmology

We can recognise five broad phases of development in cosmology, each one dominated by the ideas of certain key figures. These ideas keep recurring in later discussions because they enshrine great truths about Nature and its laws. In the first phase the foundations were laid by Galileo. He removed cosmology from the dominant influence of Christian fundamentalism, a mixture of Hebrew monotheism and Greek metaphysical philosophy. The key figures in this early period were Galileo (1632), Newton (1665–7) and – after a long interval – Thomas Young (1802), who proved that light was a wave motion. Lord Kelvin (1851) was one of several who proposed the second law of thermodynamics: that any ordered system gives way to randomly accumulated changes which end by destroying it. In 1887, Michelson and Morley demonstrated that 'the ether', believed to be the medium through which light passed, didn't exist. Ernst Mach (1897) introduced a new sophistication in the structuring of scientific theory. In 1897, J.J. Thomson discovered the electron, with hitherto unimaginable results in all our lives. The following year, Marie and Pierre Curie isolated the element radium. These discoveries and insights all provided an essential basis for the most remarkable developments in our understanding of the origins and workings of the universe.

The second phase lasted from 1905 until 1927. It saw the revolutionary transformation of cosmology by Albert Einstein. He introduced the new era in 1905 with his theory of relativity, which deals with the effects of the motion of an observer on his observations. Ernest Rutherford provided a new model of the atom (from the 'plum-pudding' atom of Kelvin to the 'planetary-type' model of today); the raisin-studded pudding of Kelvin was replaced by a central mass of the nucleus circled by electrons (1911). Using this model, the Dane Niels Bohr laid a basis for understanding the nature of spectra (1913). This was a lasting triumph of the Copenhagen School.

The third phase lasted from 1927 to 1947. It consisted in the elaboration of detail and in novel views. Among these was the theory of the younger Prince de Broglie on the nature of light as a matter–wave function (1924). A year later, Wolfgang Pauli put forward his 'exclusion' principle, asserting that there were certain rules for selection in choosing subatomic particles for mapping out atomic structures. In 1926 the German scientist Max Born advanced the notion that matter could only be known in terms of a probability function, and that therefore particle physics should be considered a branch of statistics. The principles of uncertainty (Werner Heisenberg) and complementarity (Bohr) were put forward in 1927. The former states that observation of atoms defeats objective measurement; the latter that both particle and wave phenomena are found in atomic effects where they are mutually exclusive.

The fourth phase, between 1947 and 1977, was dominated by the search for subatomic particles. High-speed particle accelerators were used to bombard nuclei and examine the fission products. Paul Dirac made a significant contribution by recognising the possible existence of 'antimatter'. Murray Gell-Mann (1964) provided an orderly classification system, and as a result he predicted that certain particles had still to be discovered – as indeed they were, a few years later.

We are now in the middle of a new phase which could best be described as 'revisionism', since it reads like a replay of the Bohr–Einstein controversy of the late 1920s. It centres not so much on cosmological questions as on debates about metaphysical issues: the nature of causality, whether observation of subatomic particles is really possible, and whether reality is merely a construction of metaphysical fantasy. J.S. Bell's theorem (1964) suggests that the principle of local causality is an illusion: signals are sent over many millions of kilometres, without awareness or intention, and are received instantaneously. This places the nature of causality in question. D. Bohm has raised similar doubts (1970) about our ability to understand the ultimate nature of reality, human existence and consciousness. But, as Galileo might say of cosmology were he here, *'Eppur si muove'*, 'Still it advances'.

Part I

The Scientific Foundations

Chapter 1

The Origins of Cosmology

Theorists are, after all, parasites. Without our experimental friends to do the real work, we might as well be mathematicians or philosophers.

H.M. Georgi

Twentieth-century lift-off

The twentieth century is a turning-point in the study of the universe and in the understanding of science itself. Among the leading innovators, pride of place goes to Albert Einstein (1905). On the foundations laid by Ernst Mach (1890) and Max Planck (1900), Einstein made cosmology the most conspicuous and exciting science of the new era. Although there have been hundreds of scientists, including many of their contemporaries, without whom cosmology would have stayed in the theologian's or philosopher's study, these three must be singled out as the heralds of the new age.

Following on their heels came the pioneers of the new age, notably the American astronomers Vesto Slipher and Edwin Hubble. Early in the century they used the first extra-large telescopes to demonstrate the expansion of the universe and to point the way to a time-scale for its existence. This led to an explosion of interest in the early history of the universe.

The atomic bomb dropped on Hiroshima (1945) was an unmitigated (and morally reprehensible) disaster. But it also ushered in the modern atomic age and encouraged scientists to take their noses out of their test-tubes and exercise some creative thinking about the morality and direction of their research. The man-made 'Big Bang' over Japan, along with Hubble's telescope, alerted cosmologists to the possibility that they had produced the first sketch of the 'Big Picture' – the 'day' of Creation. Science was mobilised for active service, and cosmology took its place as a new defence industry. Vast sums of money were set aside (and still more were promised), to be spent on making and studying subatomic particles, on trips around the cosmos (space travel for scientists), and on the purchase of essential hardware.

3

The intention of this book is to indicate the key ideas which laid the foundation for the advances made in our dazzling twentieth century. We will seek to straighten the record, explaining who did what, and when. To achieve this necessitates a highly compressed history of discoveries, not for its own sake but to clarify the methods of work and thought. Our interest is both in the discoveries themselves and in the method of thinking which led to them. This survey will include value-judgements as well as facts, so as to humanise the subject and demonstrate its value for humankind. The aim is to present the rather involved content of modern cosmology in a language and imagery that the non-specialist can understand. Numbers and equations have been avoided, and the use of symbols has been kept to a bare minimum (none at all, whenever possible) in presenting the arguments and their implications.

Some of the views I express run counter to the tradition of being polite about the ancient Greeks. Plato, for example, was in reality a rather meagre theologian and prose-poet who originated the anti-scientific campaign which dominated Western European thought not only in the Dark Ages but for centuries afterwards. Such views are opposed to the accepted hagiography found in standard histories of science and civilisation. In fact, the imposition of Greek metaphysics combined with the Roman legacy of a dislike for creative thought has been, and still is, disastrous for Western Europe. It prepared the way for the intellectual monopoly exercised by Roman Christianity, which held back the development of cosmology for many centuries and still contaminates contemporary thought.

To emphasise this point, the work of the ancient Chinese can be compared with the views of the Jesuit mission to China in the sixteenth and seventeenth centuries. This may seem an arbitrary choice, but this chapter is intended as an object lesson, showing briefly, and at less than arm's length, two things: the significance of the history of cosmology and its meaning in the contemporary scene; and examples of the arguments which crippled the progress of cosmology, and of science generally, in Eastern and Western Europe. The tradition of infallible authority attributed to ancient (Greek) thinkers held back developments for centuries. Cosmology is an especially sensitive area of science, so that alternative views were censored, and the struggle of ideas (from which truth emerges) was suppressed. Freedom of interpretation, even if scholars had shown any desire to exercise it, was not an available option.

But knowledge, unlike institutional religion, can advance only through free discussion of the alternatives. So-called 'fundamentalists' of every stripe show a misplaced devotion to the letter of scriptures as an authority on every matter. Such devotion, and the muzzling of disagreement, for a long time held back our understanding of the universe and is still doing so. Science in general, and cosmology in particular, never benefits from having its theories monitored by an overarching authority.

Cosmology and early societies

From the beginning of human history, the world of ideas has taken its problems from ordinary people and everyday life. To begin with, the questions probably bothered some individual but they quickly assumed a family and social form. Our earliest ancestors hunted, gathered and fished to feed themselves and their families. Reflecting on their condition, the thoughtful ones created a survival culture which was related to their activities and lifestyle and oriented primarily towards the means of subsistence, reproduction and survival of the group. The problems, and some answers, broadened out to become what is called the totemic view. This dealt with the universe of living things – animals, plants, and sometimes the wind, rain and environmental phenomena – and their common origins. Thus an elaborate system of relations was perceived and uniformly accepted across tribal groups.

In their most general form, the questions and answers of these early peoples amounted to a cosmological theory. Today we know the answers to many of the questions, but the details of some are still a puzzle. For example, where did animals, including humans, come from? Why does day differ so much from night? What is light 'made of'? And, the key question, how can we discover, and how can we test those answers to prove that they are correct responses to questions of origins and the nature of differences?

These and other questions concerned a few tribal experts who established agendas and rules of discussion. They might even have tried, with the limited means at their disposal, to test what was true and what was rumour. Soon these individual experts were replaced by self-selected groups, devoted to understanding the deep problems that had brought them into being. The arguments were conducted at a level appropriate to the problems of everyday living.

Obviously the participants lacked cogent facts but, more importantly, they had no method for dealing with the facts already known. Probably the select groups were committed to passing on, to relatives and friends, their new understandings, their new ways of looking at things and their discoveries. Thus the demands of their vocation were spelled out: it was a matter of informal conversations, and probably of quasi-religious rituals with occasional reporting sessions, for example when key decisions had to be made by the community. (These decisions might concern, say, questions of totem practices, or the group's relation to living-space, to Nature, and to each other.)

In return, such discussions secured the stamp of community approval. Members of the small groups, mostly ones or twos, were excused 'community' work, provided that the labour time was spent on this other, social, purpose. There are two important points here. First, these 'thinkers', were an integral, organic component of the community. Second, as a group, they had to share their thoughts with ordinary people, who also had a social responsibility to contribute to group understanding. We can see evidence for the survival of these processes in North American Indian confederations and in early groups in Australia and Africa. We are speaking, of course, about the prehistory of science, the growth of logic and the religion of Nature.

Eventually, individuals came to occupy socially defined roles as chiefs or 'guardians of the tradition', as philosophers or 'thinking-and-influence' people, as scientists or 'know-how' people, and so on. These were the 'learned professions' in embryo, perceived as emerging social roles – if they had names – and they have not survived except in unflattering terms coined by soldiers and missionaries. The first cosmologists soon split into two groups. One group claimed to know the answers (which were status-enhancing in form), and maintained that their answers were eternal by their very nature. The other group was opposed to the glib answer, claiming they needed to look for the solution, since it was the questions, not the answers, which were eternal.

This may seem all very contrived, but there was no official organisation, no rules, nothing more than conventions of good tribal manners about the method and limits of discussion – and it all grew up from there. Most importantly, there was no outside direction: everything was decided by consensus, not by majority vote. In its beginnings, thinking was a spontaneous venture, engaged in by all reflective members of the community. Structures, training, and vested interests presented as 'scholarship' came in later, corrupting the work of these cadres and distorting the process.

The origin of the sciences can be described in a similar way. In the first place there were sciences which concerned everyone, to do with the Earth and the seasons, the night sky, the planets and stars, and the general environment. In fact, the whole universe, as it was then known, was the subject of inquiry. This was the origin of the science of cosmology. The word 'science' in Latin originally meant knowledge (*scio* means 'I know from personal experience', that is, 'because I have found out'). Science was thus singled out as being based on individual, tested knowledge. Soon it became even more narrowly defined, as knowledge based on a special method. Not all knowledge fulfils the essential criteria, that is, it is not all discovery by personal inquiry, using a legitimate method. Some 'knowledge' is merely conventional 'wisdom', and often no more than fashionable mistaken ideas.

Initially, the views expressed in discussion could be criticised by any member of the 'nation' in good standing. The intervention had only to conform to the rules of ordinary, logical thought, as they appeared to the great majority. There was no censorship of opinions, so long as they were task-oriented, nor any legitimate authority other than the consensus, summed up often by the elected chiefs of the community and unanimously accepted. (Scientific conferences today seek to act on this model and, more often than not, succeed.) A tradition was established which cherished earlier knowledge, so long as it was true, or at least credible to common sense. At this stage, the only authority was truth itself, whose credibility was judged by the ordinary person, and by social consensus. Problems were considered solved when a consensus, freely attained, was reached.

These norms were crystallised by the earliest nations. They are still the norm in some aboriginal societies, such as the bush-people of South Africa, the Australian Aborigines, and those Canadian Indian groups which survived contact with civilisation with some of their tribal values intact. The village community was the norm, organised along the lines of the Mir (the pre-revolutionary Russian peasant village where a unanimous consensual group of peasant households took the decisions). In the Mir, the peasant system of native law was enforced by appointed officers. Communal control of land distribution and communal responsibility for State taxes survived in tsarist Russia until 1908, when this peasant 'parliament' was destroyed by Stolypin's land 'reform', and communal values were replaced by legal statutes.

Prehistoric science

Totemism was a consensus of this kind. Like the first cosmological theory, it was unhindered by dogma or orthodoxy. Its many different forms have been studied by missionaries and anthropologists throughout the world. It touches on a great number of issues of paramount significance for human beings. Although it answers many problems (not always correctly), it is primarily concerned with the universe of human and animal relations, setting norms for relations between, and within, these groups. Totemism enshrines a view of the universe best described as 'organic naturalism' rather than the more usual 'animism' (which suggests that it is merely a state of mind and so primitive that it is hardly intellectual). Totemism is a theory which prescribes action to solve the problems of family relationships, and which gives rules for survival to serve the primary, basic needs of humans – food and sex.

With no reference to God as its centrepiece, totemism was quickly rejected by the 'creationist' school. For millennia these fundamentalists succeeded in dominating human thought and they are still doing so. They have replaced the idea of totemism with the notion that human beings, although belonging to a vast system of human inter-relationships, are the creatures of, and therefore subordinate to, an active demiurge or Creator, who expects – and, according to some, *demands* – our worship and our intellectual and moral subservience. Theistic groups claim to detect an infinite distance between humans and the Godhead, much like the division they perceive between animals and humans. They think that the universe 'works' on a small number of non-interacting 'levels'. The model is that of the European kingdoms and principalities of the feudal age, when duchesses might dance with counts but not with kangaroos, nor even with ordinary humans, to preserve the order and law of life.

By contrast, totemism was designed to explain the natural relations within and between groups, especially between mature individuals. Passing from childhood to maturity was an event signalled by rites of passage which had to be endured as part of initiation. Totemism had survival value. It operated as an economic device to ensure the survival of the group. The children were the weakest members but also the hope of the future. They were permitted to eat anything available, particularly in hard times, whereas adults were rationed, and punished if they violated the totem rules about forbidden food. After being assigned to their totemic group they had to abstain from killing or eating their totem animal or plant.

There were also rules governing sexual relationships and marriage. These operated regardless of 'rank'; like scientific laws they were exceptionless. To some degree they controlled, and even prevented, incest and the inbreeding fatal to small groups. It is no accident that totemism and exogamy (marriage outside the group) are close partners.

Totemism is of interest to our discussion for several reasons. It is modelled on an understanding of how people actually live, and insofar as it deduces connections between causes and effects it prefigures the science of the universe, and the place of humanity in it. It is also perhaps the earliest statement in the history of our species of a proto-philosophy congenial to modern science. It has the three elements essential to science: it is coherent, it is based on experience, and its rules or laws are universal. It lacks only the scientific principle of self-correction through formal observation and systematic experiment. This is because totemism, albeit based on holistic science, was invented as a system of laws.

As implemented by Australian and other aboriginals, this theory of the living universe has to do with kinship – in other words, with the natural connections between different sections of the tribe, males and females, and their relationship to the animal, plant and fish kingdoms. The aboriginal accepts the universe and all living things – animals and humans, along with the environment – as an undivided unity. It is a complex system where each element embodies the symbols and values of every other part. Any element, whether an individual, a human group, the mountain system of the Cordillera, a mountain stream or a pack of wolves, embodies or represents all the other parts, as it does the whole. Thus, humankind represents the whole complex of reality by manifesting the same forms and same organisation as pervade every living thing. By this theory, humans, after many years of walkabouts, other communal wanderings, or of just living, may spend the rest of their time on Earth entering into some element of the landscape – a tree, a plant, a rock, a mountain or maybe a kangaroo or butterfly.

In the totemistic view, there are no absolute, eternal divisions in Nature. There are no social ranks. There is no split between the sacred and the everyday. In its entirety, and its several parts, the universe is a moral and social order, compatible with human needs and values. The theme is social solidarity, not hierarchy or vested interest. There is an established and durable link between the Aborigine and the ancestors of the tribe, as well as with certain mythic figures.

Totemism is about kinship; in the practical world, it deals with relations between people, between the sexes, between humans and the rest of the natural world. In the totemic community, kinship places each person, without discussion or right of appeal, in family relationships with other humans and animals. Each individual learns about connections with whole groups of people who are never even seen. These relationships are revealed during initiation, when the newly defined persons are assigned, and informed of, their 'places' in the great chain of being. It is a theory which classifies the whole universe of human relationships.

There are many ways in which we could relate the totemic system to the physical science of the universe, but suffice it to say that this early system is a model of the scientific system. Cosmology can be better understood if this earliest model of the universe is borne in mind.

The Chinese philosophy of science

Chinese cosmology, at least until the Jesuit missions of the late sixteenth and early seventeenth centuries, borrowed almost nothing from the astronomy and cosmology of Europe and gave back about the same. The Chinese influence on Europeans was tremendous in other ways, but it was mediated through the Babylonians, the Arabs and other Asiatic scientists. Europeans, and other foreigners, were apt in borrowing scientific inventions and processes from China, but these were not general principles or basic ideas so much as practical inventions and techniques. (They included the magnet, the compass and seamanship, as well as explosives and the investment of cities, and some rather advanced astronomical instruments.) The borrowings were, more often than not, indirectly from those nations mentioned above, so European ignorance about the sources may be excused.

Chinese physics differed in a number of ways from the science developed in the Middle East by Greeks and Arabs, though there were also some similarities. The Chinese Mohists, for example, who leaned towards atomism, worked with points and small intervals of time rather in the style of Newton's physics. Indeed, their leader Mo Ching put forward laws of motion similar to Newton's as early as the fifth century BC. However, throughout this early period, Chinese physics was dominated by the idea of waves, not atoms or discrete substances. Chinese thinkers avoided such explanations, as being out of harmony with their organic view of the universe and natural systems.

This is shown in some detail in the Chinese theory of heat. The Western notion of heat was that it was a *substance*, a weightless fluid called 'caloric' (probably thought to be similar to fire itself), which flowed into warming bodies and drained out of cooling ones. It took until the middle of the nineteenth century for European physicists to recognise that heat was not a substance but a form of energy amenable to scientific rules and methods. Caloric suffered the fate T.H. Huxley once said ultimately befalls all 'false scientific theories' – to be killed by an awkward fact, in this case that the fantasy contradicted the facts uncovered by experiment. The Chinese knew the dangers of hypothetical constructs such as caloric and phlogiston (supposed by Europeans to be given off like colourless smoke during burning). They reached that stage at least twenty centuries before the West agreed that these substances simply did not exist.

The most fundamental Chinese philosophy of science started from the concept of the Yin and the Yang, thought of as opposing organisations of power (and known to science as 'fields of force'). Yin and Yang were the whole basis of the universe and explained all change. In the West the concept of 'a field of force' had to wait until Michael Faraday's masterly research on electromagnetism in 1832. Unlike the Chinese, whose idea of Yin and Yang was philosophical, Faraday proved the real existence of a 'field' by experiment. At about the same time in Germany, Hegel developed his organic, dialectical explanation of change as driven by the conflict between the existent ('thesis') and its opposite, coming into being ('antithesis'), their subsequent interpenetration and fusion emerging as a 'synthesis'. (This is reminiscent of Dirac's matter and antimatter, see page 124.) Marx adopted Hegel's concept as the centre of his philosophy, with matter not spirit as the universal principle.

The dialectical model (whether Marxist or Hegelian) was similar but not the same as the original Chinese philosophy. It still showed signs of its origins in Western atomic theory. Also, the Chinese laid greater stress on the normal cooperation between polar opposites, implying that a process of mutual adjustment was the definitive mode of change and development. In their version of the conflict between a wave theory and particle theory of light and matter, the Chinese adopted the attitude, 'waves *and* corpuscles, please', these two forms merging in the general pattern of reality and change. Isaac Newton had much the same meld of waves and corpuscles in his concept of the nature of light, though this fact seemed to be overlooked in the struggle of ideas.

The waxing of the Yin and the concomitant waning of the Yang, and vice versa, gives rise to a wavelike, cyclical motion in nature and in human affairs. These intrinsic rhythms are seen in natural bodies in the heavens and on Earth. For example, some Chinese explained that eclipses happen because the Sun and Moon have their own Yin and Yang rhythms of brightness and darkness. In fact, the Chinese theory seems to be saying, it happens because it happens. However, there were other kinds of Chinese explanation which, while upholding the general philosophy of the Yin and the Yang, did grapple with the facts and use them to test alternative scientific theories.

The Chinese approach to the so-called 'elements' is in sharp contrast to the Greek list of four (or five) primary constituents of the universe – identified as earth, air, fire and water (and maybe a fifth element, the 'quintessence', a kind of distillation of the other four). 'Fire' is listed in the Great (Chinese) Pharmacopoeia of 1596 as a kind of element but is further analysed into various kinds:

Nature of fire	Yang type	Yin type
1 Heavenly	heat of the Sun, light of the stars, light from meteors	dragon fire, lightning
2 Earthly	fire from friction, sparks from stone or metal	burning petrol or methane
3 Human	nourishing heat, meditation heat	visceral, metabolic heat
4 Unclassified	burning marsh gas, glitter of gold, gemstones, silver etc.	

As this table shows, the Chinese attitude is fundamentally different from the Greek. The question for the Chinese scientist–philosopher is a real one. To Greek philosophers such as Aristotle, and theologians such as Plato, by contrast, it is more like a parlour game. The Chinese solution may be wrong, but at least it leads to a great number of other questions.

The Greek solution aborts the discussion, stopping the flow of ideas and observations. It becomes what we call an 'academic' question. If we ask, 'What problem does it solve, what factual question does it answer?', all we get back is an echo, 'What question?'

Chinese cosmology

The Han dynasty (206 BC–AD 220) was not the first to sponsor science as an essential concern of the State, but it was a kind of high-water mark. By the first century AD, Chinese astronomers were using armillary spheres to help observe and track the planets and the stars. Using small, hollow bronze cylinders (sighting tubes) attached to the armillary spheres, it was possible to locate objects in the heavens; focus on the celestial object could then be kept by means of a clockwork mechanism. One such clock survives. In this way, the observer could track the star or planet in its journey through the heavens. (The sighting tube was in fact, an early version of the telescope without lenses.)

As in the twentieth century, the major source of progress in astronomy and cosmology was precision instruments. These were made by Chinese craftsmen who worked closely with the learned scholars who would be using them and discussed their purpose with them. Some Chinese theories were actually named after the instruments used to collect the data on which the theories were based.

At this time, there were three separate theories in China about the nature of the universe. We should recall that the known universe at this time (for everybody) was made up of the Earth, which continued to be the subject of detailed study by Earth scientists (at least in China), most of the planets of the solar system, the Milky Way, a few near nebulas, and some visible stars. The naked eye was the sole observing instrument, as the telescope still lay about fifteen centuries in the future.

The first theory about this 'closed universe' was called *chou pei*, that is, the Sundial or Circular Paths of Heaven theory. This was accepted for a time by Imperial astronomers in first-century China as the best picture of the cosmos. The Earth was thought to be an inverted square bowl with a rim. Overarching the Earth was the sky, which was held up in the middle by a large sky-pole running from the heavens down to the centre of the Chinese Earth. The sky was 'tucked in' under the Earth at the edges, rather like bedclothes, to provide a rigid framework. The cosmos, in fact, could be thought of as a large ceremonial tent like the ones used in Imperial receptions or military exercises.

The second theory was known as the *hsuan yeh*, or Brightness and Darkness theory. It claimed that all celestial objects floated in space (empty space, we might say), except for various pieces of celestial bric-a-brac such as stars, dust clouds and planets.

12

According to the third theory, called *hun thien*, a spherical Earth was at the centre of a spherical sky. (This guess bore a certain resemblance to the geocentric Ptolemaic system which dominated Western thinking from the second century AD till the end of the Middle Ages.) Chang Yeu, a Chinese astronomer of the first century AD, provides us with a taste of the oriental flavour of this cosmology:

> Heaven is large and Earth is small. Inside the lower part there is water. The Earth floats on this; the heavens are supported by vapour. What lies beyond these parts, no one knows. It is called the cosmos. It has no end and no bounds. The heavens are like a hen's egg; the Earth is like the yolk, it lies alone in the centre.

The Chinese made no attempt to portray planetary motions as circles or ellipses, or to give the appropriate mathematical equations. Contrary to common belief, their astronomy was developed completely free of Western influence, as is documented in Joseph Needham's many volumes of *Science and Civilisation in China*. They did make star-maps, as elsewhere, but linked the planets and stars in a 'ball-and-chain' diagram (see Figure 1). Thus the idea of constellations was common to China and the West, but the star groupings and constellation names were very different.

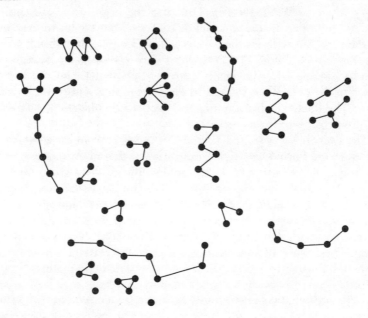

Figure 1   Ancient Chinese star maps (author's sketch from early Chinese manuscript in the British Museum)

In Chinese astronomy, the planets were not singled out as special. Stars and constellations were named rather casually after manual tools, items of observing apparatus, machines such as looms; or even leading bureaucrats in the Emperor's service. The celestial astronomer–mandarins had no interest in, or expectations of, male or female gods or stories about their alleged behaviour. This was a main difference between Greek astronomy and Chinese. The planets did not have that erotic penumbra of divine sexual exploits which has fascinated Western scholars for centuries.

Unlike the Greeks, too, the Chinese scholars were not especially interested in geometry. As astronomers would say, Chinese interests were equatorial, not ecliptic. (In other words, they studied the stars and other sky objects in horizontal planes parallel to the Equator, unlike Western scientists, who concentrated on stars and planets in the great circle known as the ecliptic, the apparent path of the Sun, Earth and other planets.)

The learned scholars in ancient China were civil servants, selected by State examinations held at regular intervals in the provincial capitals. The very top examination was held once a year in Beijing. It included mathematics, astronomy and certain other subjects, among them the writing of poetry in the classical mode. The examinations lasted a week, during which the examinees were kept incommunicado, locked away in single isolated cells, (This was the origin of our examination system, imported in the nineteenth century from China to Oxford and Cambridge, but without the locked cells.) Examinations were held within the Emperor's abode, the Forbidden City. The cell block was photographed in the late nineteenth century, but modern guides had never heard of the cells when I asked to view them on a visit to the Forbidden City. The final examination was to select the confidants and advisers to the Emperor; that is, his inner cabinet.

The Emperor was bound by the Decree of Heaven to carry out daily rituals in the Forbidden City. These ensured that no mistakes crept in to confound the celestial and agricultural routines. They also provided the monarch with a daily opportunity to petition the continuing support of Heaven for his reigning dynasty. The omens were interpreted by his ministers, astrological calculations were performed in the Bureau of Astronomy and the results were passed to those designated to receive them. These were all State secrets; officers of the astronomy department were not allowed to talk to those in other State departments.

The calendar, with the calculations necessary to ensure its correctness, was the most vital task of the civil servants. It was essential to the proper running of State affairs, the most important of which was the coordination of agricultural routines throughout the Empire: the farmers were taxed according to their yields. A new revised calendar, based on astronomical calculations, was in fact the first State document distributed to all personages by a new Emperor: its acceptance by the State bureaucracy and the powerful warlords in the provinces legitimised his accession.

This explains why the mandarins were thought to serve the Emperor best by furthering the development of astronomy and by revising the calendar. Thus earthly and celestial affairs were seen to be in step with the 'Decree of Heaven'. If these did not coincide, then warnings were sent from Heaven directly to the Emperor by signs such as comets, eclipses, conjunctions, novae (temporarily brilliant stars) and so on. As their prime duty, the civil servants read these heavenly signs during the still watches.

Clearly astronomy was of major concern to the Chinese: it had a central influence on State affairs. So the vocation of scholarship was by no means an easy option for the bureaucrats of the period. Their function was to advise, inform and give the ruler timely warnings of the condition, and signs, in the Heavens, and associated disturbances in the Empire. Scholars performed a vital State function at short notice, even being wakened in the middle of the night to do so. If they failed to inform the Emperor of approaching calamities, they risked immediate execution, along with all their relatives.

Probably the most remarkable and the best-informed Western scientist with direct experience of Chinese astronomy and cosmology at that time was Father Matteo Ricci (1552–1610). Known to the Chinese as 'The wise man from the West', he had been educated in Italy in mathematics and science, before leading the first Christian (Jesuit) mission to China from 1583 until his death.

He and a small group of Jesuit scientists, who accepted the Chinese language, culture and science on the basis that they had much to learn, were allowed to enter the Imperial realm as special guests. Ricci made himself acceptable to the Emperor and his ministers. His task was to explain and demonstrate Western cosmology and astronomical methods, especially the methods of predicting solar and lunar eclipses, which were a central concern for the Chinese.

In private letters written in 1595, Ricci singled out what he described as the 'absurdities' of the Chinese scholars he worked with. Among those he listed were the following: they believed that there were five elements, not four (they excluded air from Aristotle's list, but added metal and wood to earth, fire and water); they believed that the Earth was square and flat (there were other schools of thought, as we have seen); they thought that there was only one sky and not ten (this refers to the crystalline 'spheres' which Ptolemy believed to surround the planets); they thought that the sky was empty (void) and not solid, and that only the stars and planets were located in this void (the Chinese didn't know or understand the properties of air which, according to Western thinkers of the time, filled the space between heavenly bodies). We might add that the Chinese did not know about ether either. Such statements indicate the source of Ricci's own views: Greek cosmology, as processed by Saint Thomas Aquinas and declared to be Divine Truth by the Pope himself.

Chapter 2

Greek Cosmology and its Aftermath

Platonism

Many cosmological ideas were expressed by Greek scientists and cosmologists. Unfortunately, although we have an almost complete record of the thoughts of Socrates, Plato and Aristotle (the intellectual sources of the European Dark Ages), very few of the scientific alternatives have come down to us. We have only fragments, most often in the form of quotations reported by opponents for polemical advantage. The logic-chopping method of Aristotle and the religious ideas of Socrates and Plato (principally that thinking about the gods was the province of élite minds only, and was far above such mundane matters as factual proof or the practicalities of ordinary life) won the fight for Greek (and European) minds, overriding the influence of the Greek empirical scientists.

Greek scientists were as little disposed to accept Plato's invitation to intellectual hara-kiri as any contemporary scientist would be. Instead of studying the stars for themselves, Platonic astronomers used the heavens to discover the qualities of God that were supposedly found in Nature. Similarly, Platonic zoologists would study God's solicitude for animals and humans, and so on. If knowledge in any field, especially the scientific, failed to reveal divine teleology, then that 'knowledge' was wrong and had to be rejected. By extension, such an approach, as interpreted by Goebbels, inspired the Nazi burning of all 'Jewish' writings and – more importantly – continues to this day to energise fanatic fundamentalists.

The Greek scientists, who are almost never named by Plato, kept their own counsel in the face of such religious obscurantism. This was either because they knew nothing about Plato except as prose-poet, theologian and bear-leader for budding politicians, or because they decided that it was a waste of time to argue. Those opposed to the anti-science of Socrates and Plato were concerned with, and talked about, matter; they rarely talked about God and souls. Aristotle should be counted in the group of scientific realists since he talked in a direct factual logic about intellectual questions. He seemed to accept Plato's notion that aesthetic

17

value was a test of scientific merit, but he was less of a pagan than Plato. His basic writings are compatible with monotheism, not with polytheism. He rejected Plato's notion that the real world was inaccessible to human knowledge because it was out of bounds to human perception. Although his writings were later used in support of Christian dogmatism, he sponsored rational inquiry, and avoided the excesses of Plato. His scientific work on growth, the parts of animals, and physics were based on first-hand empirical observations and studies. He cannot be held responsible for the later misuse of his views by theologians.

Plato was the most influential spokesman against scientific knowledge. His writings take the form of reported (but imaginary) discussions, which involved Socrates as interlocutor. Socrates jusified his preoccupation with cross-examining all and sundry by asserting that the Delphic oracle was correct in singling him out as the wisest man in Athens. The punch line was, of course, that he (Socrates) was always conscious that he knew nothing, whereas all those he talked to were too ready to lay down the law about things of which they knew nothing. The Delphic oracle thus conferred on him the right (or religious duty) to argue with those who were regarded, or who thought of themselves, as having some special knowledge. This claim to superior wisdom was made repeatedly by Plato–Socrates, albeit tongue-in-cheek and for the amusement of students who were aware of the discomfiture that Socrates' opponents were caused in these imaginary dialogues. (We should recall that in these non-events, Plato's Socrates always won – he was the Perry Mason of his age.)

In one dialogue, there is a discussion about the benefits which would follow from burning all the works of the scientific thinker Democritus. This betrayal of scholarship may indeed have taken place. If so, it would explain why so few scientific and anti-Platonic materials have survived. It would also be in keeping with Plato's 'thought police' attitudes expressed in *The Republic*, his major dialogue on the 'ideal state'.

The *Timaeus* is one of the last of Plato's dialogues (380 BC or later). In it, Plato aims to declare the essential principles that account for the origin of the universe. He also tries to explain the sciences of astronomy, physics, physiology and biology in rather fine detail. The *Timaeus* is a dreadful parody of science. Its method of inquiry is metaphysics, using definition and logic unaided by empirical knowledge in a vain attempt to answer what are essentially questions of fact. And, for no good reason, he throws in the myth of Atlantis.

In his dialogue *The Republic*, Plato claims that ordinary people have souls of lead, not gold as their 'guardians' have, nor silver as the police and soldiers have. Thus, he concludes: they need to be told pleasant-sounding lies as myths (his own words) for society to remain stable. Atlantis is just such a pleasant lie. The myth has a basis in fact (the true site of 'Atlantis', and the causes of its destruction, were revealed in a geological report from 1965), but this is obscured by Plato's fantasies and his ignorance of Egyptian numbers – the true story is recorded on a standing stone in an Egyptian temple. Because of Plato's misleading account, there were reportedly more than 20,000 failed attempts to locate Atlantis.

Plato follows what is, for him, a standard detour, pointing out that empirical science (as we must now describe cosmology) cannot give us 'certain knowledge'. (What he doesn't say is that this is true of all knowledge, none of which is 'certain', except for platitudes and definitions.) Therefore, he concludes, we need pay no heed to science, because 'it changes all the time', and as mere opinion is of no interest to the Academy. Metaphorically speaking, he swears by the gods that mathematics is different, since the deductive method yields both truth and certainty. Quite so, but only if our initial assumptions are true (that is, reality-based), and our deductive process is without error.

Of course, functionally speaking, Plato was a theologian, with no standing as a scientist, or even as a mathematician, though he frequently asserted he knew the best way to advance both these subjects. He was opposed to materialism, believing that reality, as we know it, is a mere shadow-play. It is hardly surprising then that his substrate (or basis of reality) is not matter, but space (*chora*). In this connection, he also refers to *ananke*, meaning destiny or the principle of necessity. *Ananke* can be interpreted in various ways, but Plato takes it in the theological sense, that it works in human affairs to bring about the obscure, but ultimately intelligible and beneficent, purposes of God. In fact, this is probably the only cosmological principle he contributes to our ideas of how the universe functions. (The Methodist hymn puts it more succinctly: 'God moves in a mysterious way/His wonders to perform'.)

Nowadays we operate the rule that no appeal should be made to divine wisdom or foresight (that is, no teleological explanation should be considered) until all other explanations have been tested exhaustively and systematically ruled out. In other words: the appeal to divine wisdom and intervention should come, if at all, only at the end, and not at the beginning, of a scientific inquiry. Nor should any supernatural explanation be introduced until every natural causal theory has been eliminated. Moreover, the existence of God (or in Plato's case, of gods and goddesses) has to be established; it is not enough merely to assert it.

Why then, do we bother to discuss Plato at all? One reason is that for centuries philosophers, and Greek scholars, held science in thrall to his views and accepted him as a source of wisdom second only to his student Aristotle. More importantly, there has been a recent revival of his views, cosmological and others, among some mathematicians. For our purposes he serves to show that, by themselves, literary and philosophical excellence are not enough in scientific discussion, especially where cosmology and the origin of the universe are concerned. The final judgement on any scientific theory or statement must rest on the answers to two questions, is it true, and does it work *because* it is true?

Aristotle's legacy

Described by Dante as 'the Master of those who know', Aristotle for many centuries controlled the thoughts of Western scholars. In the 12th century Roman Church theologians officially decreed him the authority on logical analysis, psychology, natural philosophy and many other subjects basic to theology. Having been 'sponsored' by Saint Thomas Aquinas (1226–74), himself the ultimate authority in Catholic theology, Aristotle was a main resource in the discussion of topics in these areas well into what we might call the modern 'scientific era'. In the 17th century, for example, a Jesuit professor of mathematics denied the validity of Galileo's observation of sunspots on the grounds that, having read through Aristotle, twice, he'd been unable to find any reference to sunspots, and they therefore could not exist. Like other learned scholars of the day, he simply refused to emulate Galileo and look through a telescope.

Aristotle was a student at Plato's Academy but showed an ample degree of independent thought. Unlike Plato, he was a realist, who accepted that the external world exists independently of human thought. He was in no sense a mystic or an idealist. He worked at various sciences such as astronomy and biology and wrote several treatises on the development of animals, as well as one on the heavens, and other texts on a great variety of topics. He had a general philosophy, or method, for the study of nature (which he describes in his *Organon*). Although he lived in the fourth century BC, throughout the Middle Ages he was regarded as the universal authority on all matters, religious as well as secular. Even as late as 1750, a Papal encyclical (letter to all Catholics) bound at the front of the Douai translation of the Bible, orders that teaching in seminaries should be based on Aristotle's writings, *as interpreted by Aquinas* (see next section). Aristotle's ideas and mode of thought preceded the scientific revolution and are alien to modern thinking. But they were accepted by Western scholars almost uniformly for two thousand years or more. The few honourable exceptions, scientists and others who tried to think for themselves, were either reduced to smoke and ashes or forbidden to write or speak to other scholars. The devil was thought to be very active in promoting heresy – a belief that cast a spell over people's minds. Indeed the spell was broken only when Francis Bacon (1561–1626) in England, and René Descartes (1596–1650) in France, revolted against the whole Aristotelian system, especially the *Organon*. Aristotle cannot, of course, be blamed for being adopted as the Roman Christian oracle many years after his death, nor for the sins and stupidities of the Inquisition. He was seen as essential in filling the intellectual gap identified by Saint Paul in the 'faith of the fishermen'.

Although Aristotle immersed himself in scientific matters, he was much influenced by the thoughts of his master, Plato, most obviously in cosmology and astronomy. Aristotle and Ptolemy were the source of the items in Father Ricci's catalogue of Chinese 'mistakes' (see page 15). Today we would regard most of Ricci's Aristotelian fantasies as nonsense, since astronomy has long accepted the views of the Chinese mandarins. We would only take exception to the Chinese view that the Earth is square and flat. We know it is spherical, and can so testify, having seen pictures of it from outer space. But, of course, in Imperial China, unlike Christian Europe, it was possible to belong to one of many schools of thought. This was the period in China when 'a thousand flowers blossomed'.

Aristotle (as interpreted at the time of Galileo) believed that everything in Nature had its 'proper place' and that all bodies in the Heavens and on Earth moved by 'natural' motion to occupy those places. Also, the straight line and circle were deemed superior to all other figures, being more aesthetically pleasing and 'perfect' in the sense of being complete. Therefore bodies moved 'naturally' on earth in straight lines, but in circles through the heavens. Thus, bodies actively sought and found their natural places. At this time it was accepted that, by divine ordinance, the Earth was at the centre of the universe. All other heavenly bodies, including the Sun, moved around it, each in their own crystal sphere, and all circling east to west.

Since a vacuum was impossible in Nature, there could be no 'empty' space in the universe, and so the void proposed by the Greek philosopher Democritus was an illusion – or so said Aristotle. Therefore the space between the Earth, Sun and planets must be filled with some 'medium' we cannot see (a little like air or ether). This medium was imagined as a very fine-grained material, with no taste or smell (unlike the ether anaesthetic, which was known only as a bouquet in wines). It was rather like a kind of invisible, colourless gossamer, like the cheese-cloth used in recent times at séances to represent 'spirits'. Bodies on Earth and in the Heavens were set in motion by a prime mover, which imparted a certain amount of motion to set things off. They then continued to move through the thin but resisting air–ether medium. As they did so, air rushed in behind them because, as Aristotle's disciples would repeat parrot-fashion, 'Nature abhors a vacuum'. The moving body was thus pushed on its way and continued in motion because of the continuing inrush of air. This was called 'violent motion', as opposed to 'natural motion', because it was due to interaction between the moving body and some external force. (The encounter was clearly not 'natural' because one or both of the bodies in a collision, such as air colliding with the moving body, could not have been in their natural places.)

This argument implies that if motion and rest are considered to be two conditions of matter, then rest will be superior to motion, just as the circle is superior to any irregular figure. In the same way, the scholar sitting, and maybe thinking, is superior to his domestic slave at work on household duties. Such a way of looking at the universe persisted for a very long time – from the fourth century BC to the 16th century AD and beyond – and is not so unusual even today.

The Chinese thought otherwise. Their science being more solidly based, they could shrug off Confucianism, the Chinese equivalent of Platonic–Aristotelian thinking. It remained only one of 'a hundred schools' and was adopted temporarily as a state philosophy only during the Han dynasty. The Chinese believed that action was generally superior to non-action, but they were wary of making absolute (verbal) distinctions of the Greek kind.

How then did the Chinese regard Father Ricci's cosmology, given his view of Chinese 'stupidities?' His methods of calculating the times of eclipses had great merit, they thought, and they conceded that they had much to learn from him. (This was the Ming period in China, which may have been advanced in art but was somewhat degenerate in scientific scholarship.) But, they went on, the Father was quite 'confused' about heavenly bodies. They made this judgement while Galileo's system was being weighed by the Congregation of the Holy Office for heresy and while Galileo was forbidden to teach, defend, write or even speak about the Copernican system. Father Ricci gave the Chinese the impression that he was unsure whether the Earth went round the Sun (Copernicus' heliocentric system supported by Galileo), or the Sun went round the Earth (Ptolemy's geocentric system supported by the Church). Ricci was in fact a closet Copernican, but was under the discipline of his superiors who, of course, knew nothing about astronomy except what Aristotle through Aquinas had told them. Ricci had already been in trouble for saying Mass in 'the Chinese rite'. Like the Jesuits of the period, he was thought by well-intentioned Catholic obscurantists to be too clever for his own good.

On the question of Galileo, the Jesuit father could not declare himself before the Pope had spoken. Here I am reminded of an experience of my own in 1950, when as a member of a nuclear disarmament campaign I visited religious leaders in a certain town. Having discussed with the parish priest the question of the immorality of indiscriminate war on unarmed civilians, as a 'certain question of faith and morals', we outlined our policy to him. He listened politely to our argument, that the mass killing of unarmed non-combatants such as women and children was immoral and an act of terrorism. But then he surprised us by saying he could not express an opinion 'since his superiors hadn't spoken on the issue of nuclear weapons'. There was nothing more we could say, except perhaps to point out that theologians had already condemned such killings, defined as genocide, as mortal sin. Presumably something similar was what embarrassed Father Ricci: his superiors had not yet finished with the issue.

In 1633, 23 years after Father Ricci's death, the Vatican decided permanently to forbid Galileo to speak or write about the Copernican system, and confined him under house arrest. The Copernican theory was declared, on divine authority, to be false. Galileo was accused and found guilty of heresy, even though the new theory was never formally denounced as heretical. In 1950 some three centuries later, an encyclical declared the Ptolemaic theory 'naive', though suited to the understandings of a simpler age; it was not until 1992 that the Vatican finally announced that Galileo had been right all along. One can understand and sympathise with Father Ricci's silence: unlike Galileo he had taken the vow of obedience.

Aristotle's works were used to silence, if not to confound, Copernicus. The main reason for doing so was that the Copernican system could not adopt the value-laden frame of reference inherited from the Greeks Pythagoras and Plato. According to this, good circles and straight lines were better than ugly quadrilaterals and curved, but unfinished, non-circular lines. Specifically, the Copernican theory of a mobile Earth circling the Sun was contrary to the teaching of Plato and Aristotle. There was little difference between them in astronomical calculations because Copernicus' model was only slightly less complex than Ptolemy's. Both theories were still 'Earth-bound'; they worked equally well in all calculations of relative motion. Galileo demonstrated that either the Sun or the Earth could be taken as the moving body to serve as the point of reference, but when speaking of the solar system there was an advantage in taking the Sun as the centre, rather than the Earth. This re-centring of our picture of the universe is sometimes known as 'Galileo's relativity theory'.

The medieval flavour of scientific discussion is captured in the following quotation from Galileo's *Dialogue Concerning the Two Great World Systems* (published in Leyden, Holland in 1632 and later placed on the Church's Index of Forbidden Books). Simplicius, the Aristotelian representative in the dialogue, is refuting Copernicus' model of the heavens and the problem of a mobile Earth:

To disprove this idea (that the Earth moves), Aristotle's axiom is in my judgement most sufficient. He says that in the case of a simple body such as the Earth, only one sole, simple, motion can be natural. But here, in this case [Copernican theory], the Earth, which is a simple body, is assigned three or maybe even four, motions, all very different from each other.

Simplicius then goes on to list the Copernican motions as follows: the motion of a heavy body (the Earth) towards its natural centre (that is, the centre of the universe); a circular motion around the Sun; the rotation of the Earth about its own centre every 24 hours.

Aristotle's influence on Western science was just as harmful, and in fact more so, than Plato's clearly expressed opposition to 'scientific' thought. Aristotle's method was supposed to encourage people to consider facts rather than opinions, and unlike Plato, Aristotle was not always seeking the most inappropriate teleological excuses. But he could have spared us a lot of trouble by stating, or at least hinting, that deductive logic should not be allowed to obliterate empirical fact – that logic should always be checked by experience, which was a primary guide to truth. (However, as we said earlier, Aristotle can hardly be blamed for the stupidities and dogmatism that kept European nations in thrall for centuries to his 'axioms' and his 'opinions', however tenuously expressed.)

The healthy scientific tradition in ancient Greece, so different from Platonic and Aristotelian ideas, originated with Heraclitus, but developed as a philosophy through the Stoics of the third and fourth centuries BC. They taught that the cosmos was a unity and essentially rational, and that matter should be treated as a whole and regarded as continuous, not atomistic and discrete. The world was spherical, they said, and could be understood by humans. The teaching left room for divine providence, but not such as to interfere with their thinking. Indeed their general philosophy was deterministic, but unlike Plato they regarded human beings as free agents who can become aware of the true nature of things. It was an undogmatic realism, and, like science, it was eclectic.

Ptolemy's system
The most important contribution of the Greeks to cosmology was made by the Greek-Egyptian mathematician and geographer, Ptolemy. We know almost nothing about Ptolemy himself except that he lived in Alexandria between about AD 127 and 145 and taught at the great Alexandrian library, which was also a university. He wrote an account of a geocentric universe (one in which the Earth is at the centre) in a classic text with the Arabic title *Almagest* (which means 'the Greatest'). Ptolemy proposed a model for the part of the universe then known to

humans, the relatively small area visible to the naked eye. The Ptolemaic system was accepted by Christian Europe because it was compatible with both Aristotle and the Hebrew Bible. It provided a cosmology for the 'closed universe' – a kind of one-eyed picture acceptable to common sense, and even to Platonists and Aristotelians. A non-moving Earth was at the centre, with the other planets revolving in circular fixed paths around it. It seemed so obvious to humans on Earth that this was the way things were!

In fact, Ptolemy's model was more complicated than this. The Sun circled a stationary Earth which, in turn, was circled by the Moon. The Earth was also circled by the planets belonging to what we now call the

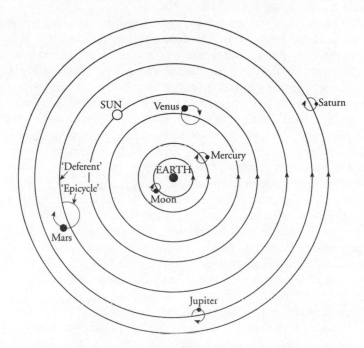

Figure 2   The Ptolemaic system

Ptolemy's solar system had the Earth at the centre (geocentric) with the Sun and the remaining planets travelling on circular orbits around it. In order to agree precisely with the observations, Ptolemy introduced a system of *epicycles* and *deferents*, and by carefully scaling the sizes of the epicycles to the size of the deferents he was able to reproduce the observed planetary motions with an accuracy of 1%. The clockwork nature of the Ptolemaic system along with the ideal of a geocentric universe appealed to many and it would be more than 1300 years before a Sun-centred (heliocentric) system superseded it.

solar system. Each planet circled the Earth at a set distance from it. Unlike the stars, the planets 'wandered' about the sky, but in fixed paths, returning at rather precise intervals in their individual cycles to retrace the path just run, with only minor variations. Each planet moved in its own orbit, at different distances and at different speeds. (Chapter 3 gives up-to-date measures and other information on the solar system.)

Ptolemy based his system on five basic assumptions: 1) heaven is a sphere which rotates; 2) the Earth is also a sphere but is at rest; 3) the Earth is the centre of the heavenly sphere; 4) the ether encloses all heavenly bodies; 5) the ether too is a sphere revolving in unchanging circles, and each planet revolves in its own crystal sphere, which surrounds and separates it from other planets.

This model was accepted for many centuries until some unsolved problems demanded, and then forced, a reconsideration. The main issue was whether the Earth, not the Sun, really was the centre and focus of the solar system. Galileo's use of the telescope for observing the heavens made those who were interested aware that matters in the real world did not conform to the aesthetics that the Greeks demanded. The Sun was not 'perfect' but had spots on it, which were visible even to the naked eye. This discovery was quite contrary to accepted wisdom. It suggested that perhaps other problems of the universe might also have a logical basis different from that proposed by Aristotle.

This shift of focus, or more properly change of paradigm, came to be known as the Copernican revolution. Copernicus himself, a minor canon of Cracow cathedral in Poland, died in 1543, on the same day as the printer published his new theory of the heavens. Galileo thus found himself in the forefront of the battle, rather like Uriah (Bathsheba's husband in the Bible story of David and Bathsheba), not knowing quite how it came to pass but willing to lead the troops into battle.

Physical science in pre-modern Europe

The French Revolution of 1789 signalled the true beginning of the modern age. Just as centuries-old dynastic interests were suddenly assailed on all sides by democratic aspirations, so there was a democratisation of thought: ideas were no longer the possession of a privileged and autocratic élite, blinkered by centuries of scholasticism and religious dogma, but were available to all, and of universal interest. Voltaire, for example, and his mistress Madame du Châtelet, who translated Newton, became leading European authorities on the new physics.

At this time, scientific questions had not yet been confused with national prestige. The question of priority, that is, who was first to make a given scientific discovery, confuses the issue by diverting attention to national prestige. Arguments over priority nurture chauvinism (in its original sense, of a rather mindless nationalism), leading to bad relations between scientists, and with no real benefit to anyone except the scandal-mongers.

In the period before the French Revolution, differences were regional, not national. The Christian religion, from early times, had been a significant unifying cultural force. Paul's Christianity, a drastic revision of the original message, emerged as an intellectual system from a synthesis of Greek pagan philosophy and Hebrew monotheism and Messianism. The merging of the two intellectual currents was monitored in the West by Christian theology, and developed as the new philosophy of scholasticism in universities and seminaries.

At the time of the Reformation, one of the reactions of the Vatican was to establish its censorship bureau, and the Index of Prohibited books was a major part of its counter-Reformation strategy. The Index was designed to supervise and direct thought and expression in conformity with the scriptures and Christian doctrine. The aim was to prevent further heretical movements from developing. The Jesuits aimed to produce the same effect, not by persecution and censorship, but by education through their missions at home and abroad. They carried on a bloodless battle against intolerance and Protestant persecution. Such was the framework within which science developed in the West, uneasily dependent on organised religion for its cadres on the one hand and borrowing from Oriental science and learning on the other (usually without acknowledgement of the debt since this could lead to a different kind of problem).

The destruction of the great library at Alexandria and the movement of scholars westwards along with a quantity of manuscripts, scrolls and texts, had brought about a renewal of academic learning in Western Europe. The spread of Roman garrisons and Roman religions across Europe had done nothing for the culture and education of the indigenous peoples. The Romans were a practical people, whose excellence lay in engineering technology, especially military. They made few advances in theoretical science and almost none in cosmology. Apart from perhaps a few Christian monks in their monasteries, or some cultured clerics in the ecclesiastical hierarchy (who knew a little Church Latin but no Greek and certainly no Chinese or Arabic), the intellectual classes possessed little in the way of scientific knowledge. Most people, as today, had their attention focused on other matters, like bread and circuses.

The international language in the West was Latin; in the Orthodox Eastern Churches it was a kind of pidgin broken Greek (for example, Church Slavonic in the Russian Church). The Chinese and Arabic languages (which covered most of the real science in the Middle Ages) and the discoveries that they dealt with were largely unknown to Western scholars. Where a Latin translation of a research report, or a Latin compendium of results, was lacking, scientific writings were unknown territory even to the professionals in the educated class. As to the Orthodox Churches, they were not merely indifferent but actively hostile to secular knowledge. So too were the religious leaders of the Protestant sects. These were indeed the 'Dark Ages' for scientific work in Europe, both eastern and western. Indeed, apart from the creation myth in the book of Genesis, there was little that could be described as cosmology in Europe in these times.

By contrast, today we have a precise, factual knowledge of the universe – by no means complete, but established with the same degree of certainty as the solutions to other physical problems. Moreover, as a general principle, speculations are not ruled out in scientific matters – not even the wilder speculations of some competent scientists. But they are subject, at a very early stage, to the discipline of logic and the test of objective reference, before they can be provisionally accepted as probabilities. At a certain high level of probability they may become part of the scientific paradigm. In this regard, the rules of scientific reasoning are as precise as those informing legal decisions at the highest level, though they are less well known.

Chapter 3

The Take-Over by Physics

The history of science teaches that the subjective, scientific philosophies of individuals are constantly being corrected and obscured and . . . only the very strongest features of the greatest men are, after some lapse of time, recognisable.

Ernst Mach

Cosmology as an exact science

Greek 'science', in the true meaning of the word, was the work of a few first-class minds, thinkers and inventors such as Eratosthenes, Archimedes and Democritus. The Greek class system resembled the slave society of the Confederate States of America before the Civil War. Women were excluded from serious business, and slaves were without a voice or means of expressing an opinion. In contrast to trade, politics and the law, science took a back seat. Theoretical and applied physics were quite distinct from the metaphysical speculations of commentators. The 'action' of physicists (to be contrasted with the 'comment' on their action by metaphysicians), was still only concerned with statics, that is, with physical bodies in a state of rest, such as ladders leaning against walls, or weights hanging from pulleys. The only movements studied (for example by Archimedes, 287–212 BC, in Syracuse) were the lifting of loads by levers, or by block and tackle. Very probably, Greek stupidities in philosophy inhibited the study of self-maintaining motion: some philosophers denied the very existence of motion while others concealed it by a logical sleight of hand. According to the victors in the battle of opinions, knowledge was unrelated to experience. If it contradicted experience directly, or denied its relevance as a principle, then this was seen as a clearly superior brand of truth.

Metaphysical discussion, in this *a priori* mode, claimed that motion did not exist. On this question, Aristotle had made a claim for metaphysics in opposition to science. His verbal analyses (quite inappropriate to the matter in hand) were aimed at establishing the existence of different kinds of 'movers', including a prime mover. During the Dark Ages, when Greek philosophy was mediated by Roman Christians, Christian scholastics converted the Aristotelian 'prime mover' into God Almighty simply by the use of capital letters. So doing, they claimed cosmology, from early Christian times throughout the Middle Ages, as a branch of theology, and therefore not amenable to correction by scientific facts. Logic in the *a priori* mode (that is, the analysis of phenomena by logic without any check by direct experience, or even in defiance of it) precluded the study of dynamics. From reading the dialogues of Plato it is obvious that 'science', in any modern sense, was a dirty word for Greek conservatives. Aristotle failed to comment on this obscurantist attack, and in some ways he even accepted it. Significantly, it was these two among the Greeks who were singled out to provide the intellectual content and social norms for Christian thinkers.

The science of dynamics: the inside track

Galileo Galilei

Galileo (1564–1642) broke the Greek stranglehold on physical science and established dynamics as a new science providing us with exact knowledge of moving bodies. He also made a detailed analysis, based on his experiments and written in dialogue style, of the stupidities of the Aristotelian metaphysics of his day. In his *Dialogue Concerning the Two Great World Systems, Ptolemaic and Copernican* (1632), he categorised his opponents as 'the most reverent and most humble slaves of Aristotle, [who] would deny all the experiences and observations in the world, and would even refuse to look at them . . . they would say that the universe remains just as Aristotle has written about it'.

Galileo was the first to turn his new-made telescope on the heavens. In relatively short order, he discovered and estimated the heights of the mountains of the Moon, proving that the Moon was a solid body similar to Earth (this was a matter of dispute in his time). He also discovered the satellites of Jupiter and Saturn, observed the phases of Venus and discovered spots on the Sun. But his most crucial revelation was his conclusion that there was no difference between the laws of Nature as they operate above the Moon (on the so-called 'heavenly bodies') on Earth, and beneath the Moon ('sublunary bodies'). This implied that if we could study the movements of bodies in physics laboratories here on Earth, by so doing we could discover and prove the laws of planetary motion. It also enabled scientists to construct working models of the solar system (some of which were made generally available), as well as simulations of eclipses.

Galileo's recognition that scientific laws were 'universal' represented a giant step forward for cosmology as well as for astronomy. In fact cosmology, like dynamics, can be taken to have been created by Galileo. Some would even say that it was Galileo who was the real founder of 'modern' science, rather than Newton (who remained captive to many 'ancient' Aristotelian fictions).

Galileo established that useful cosmological work could be done in a 'sublunary' laboratory. But to carry out such experiments required a method for the accurate measurement of short intervals of time. In this connection there is a famous story of Galileo observing a swinging chandelier during Mass in the Cathedral in Pisa (1583). He timed the swing by taking his pulse-beat (it was probably during the sermon when attention sometimes wanders). He discovered that the amplitude of the swing (the distance travelled) generally had no effect on the period (the time for one to-and-fro movement). The period, timed by his pulse, did not vary.

For timing experiments in the laboratory Galileo used not his pulse but a water-clock, or *clepsydra*, modified for his purpose. Using a large vessel with a small hole near the bottom, he could control the water flow by covering the hole with his finger and then removing his finger during the interval to observe the motion. He then collected and weighed the water released to measure the time taken by the moving body. (This measurement was not in seconds, but in standard units, but they were all that he needed.) By this simple means he recorded the velocities of bodies over time, whether on an inclined plane, a level surface, or in free fall. He discovered the relations between time, velocity (speed with direction), distance and acceleration.

Galileo was by no means the first to lay into Greek metaphysics. Aristotle's system was earlier subjected to a devastating attack by the mysterious figure known only as John the Grammarian, who gave a detailed mathematical and logical analysis in which he exposed its inherent contradictions and criticised its rejection of any commonsense dealings with reality. However this critique has been largely ignored. Leonardo da Vinci, as we know from his private notebooks, had carried out some experiments on the law of inertia. But any criticism of the Greeks was brushed aside by the 'learned' poets, philosophers and thinkers of the day.

Johannes Kepler
Observing planetary motion through the telescope, Johannes Kepler (1571–1630) delivered the *coup de grâce* to the Greek picture of the solar system. He demonstrated that the orbits of the planets were not even circular (heavenly motions, for Platonic orthodoxy, had to be 'perfect', therefore circular). Kepler's three laws of planetary motion (see Figure 3) were 1) the planets go round the Sun in ellipses (close to

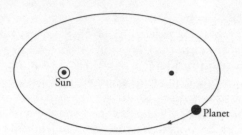

Law 1: Planetary orbits are ellipses with the Sun at one focus

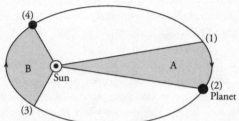

If the time taken for Planet to go from (1) – (2) = time taken to go from (3) – (4), then Area A = Area B

Law 2: Planets sweep equal areas in equal times (Area A = Area B)

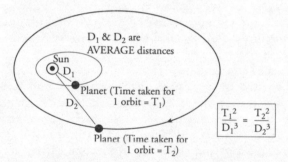

D_1 & D_2 are AVERAGE distances

Sun

D_1

Planet (Time taken for 1 orbit = T_1)

D_2

Planet (Time taken for 1 orbit = T_2)

$$\frac{T_1{}^2}{D_1{}^3} = \frac{T_2{}^2}{D_2{}^3}$$

Law 3: If D is the *average* distance of a planet from the Sun and T is the time taken for it to complete 1 orbit then the equation $\frac{T \times T}{D \times D \times D} = \frac{T^2}{D^3}$ has the same value for all planets

Figure 3   Kepler's laws of planetary motion

circles), in which one focus of each ellipse is the Sun; 2) the (imaginary) line which joins the planet to the Sun sweeps out equal areas in equal intervals of time; 3) the cube of the distance of each planet from the Sun varies as the square of the time the planet takes to complete its journey round the Sun. (We should remember that only the solar planets Earth, Mars, Jupiter, Venus, Saturn and Mercury were known at this time, Pluto and Uranus had still to be discovered; see chapter 4.)

Christiaan Huygens

Christiaan Huygens (1629–95) was a Dutch scholar who made important contributions to dynamics, advancing the work of Galileo. He invented the pendulum clock, a vital piece of laboratory equipment for measuring small intervals of time. He also produced evidence to show that light was a wave phenomenon, in opposition to the theory attributed (incorrectly) to Newton, that light could best be explained as tiny particles emitted by an illuminated body (the particulate or corpuscular theory). The controversy about the true nature of light was launched; along with the nature of gravity, it moved centre-stage again in the twentieth century when taken up by Einstein. Huygens found Newton's views unacceptable on both matters.

Huygens also discovered the fine structure of the rings of Saturn and identified the stars in the constellation of Orion. He carried out distinguished work in explaining circular motion which was essential for understanding planetary motion. The idea can be illustrated by the children's game in which a stone, on the end of a string held in the hand, is swung in a circle.

The motion can be resolved into, and explained as, a compound which includes the acceleration created by the force we apply to keep the stone from flying off. This acceleration is directed inwards to the centre, and so it continually changes direction, as is noticeable from the tension in the string. The velocity too continually varies, because the stone's direction keeps changing even though its speed remains the same. Acting together, these two forces combine to give the circular movement (see Figure 4).

This explanation can be tested in a laboratory, where the force keeping the stone in its circular path can be measured, using a spring balance tied to the string.

In this way the theory differs from the *ad hoc* reasonings of common sense, but builds on them. In a long series of experiments, Huygens disposed of several more complex problems, such as the compound pendulum, oscillatory motion and centres of gravity.

Huygens' quarrel with Newton's theory of gravity was that there was no way to explain how any force, not just gravity, could act at a distance. In common with other working scientists, and long before Ernst Mach declared it as a guiding principle, Huygens refused to accept verbal reports or descriptions which could not be checked by observation or controlled experiments. Analogy, he said, was legitimate as illustration, but not as a method of proof. Reasoning based on quantity and number, he insisted, was the only acceptable way of reporting physical relations. Statements are thus made precise and testable in different settings. But although logic is essential for correct reasoning, it is not a method of research. It lacks any power to advance, but can be used, and often is, to

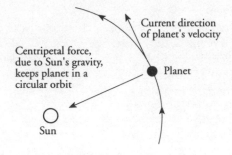

Planetary motion

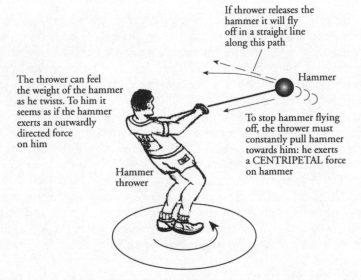

Figure 4   Centripetal and centrifugal forces

retard empirical knowledge. Precise observations provide facts to reason with.

For Newton's explanation of action at a distance, Huygens substituted his own theory that gravity was not a linear force, but a wave phenomenon, transmitted by pressures and impacts. The complex subject of gravity will be dealt with later; it is an ongoing problem, still without a solution.

The seventeenth and eighteenth centuries were a period in physics when mechanical models were widely used. The paradigm was copied from Descartes and LaMettrie, both of whom portrayed animals and humans as machines responding to sensations in a connected, machine-like 'reflex' fashion. This was a rather 'mindless' reaction to the cynical abuse of religion by princes and many Church leaders. They, who were themselves 'practical atheists' (to borrow a useful phrase of the philosopher Jacques Maritain), corrupted true religious feelings by transforming religion into a self-interested social 'pacifier' of the masses. The militant emperor Frederick the Great of Prussia clearly expressed this motivation in words and deeds alike.

In projects such as those of Huygens, the most important questions of dynamics were investigated by experiment. The central problem was to discover ways of measuring the physical variables (mass, distance, velocity, time and so on) under laboratory conditions, where systematic variations could be introduced and their effects measured. This work, carried out by members of scientific societies, and by amateur scientists, was the solid base for the great advances of cosmology in the twentieth century.

Huygens was the first to measure the value of g, that is, the number which represents the acceleration (rate of change of velocity) of a freely falling body. This acceleration, like gravity which explains it, is directed towards the centre of the Earth, unless we happen to be astronauts in space. In the Imperial (foot–pound–second) system, the speed of bodies falling unimpeded under the influence of gravity (free fall) increases by 32 feet per second every second. In the metric (centimetre–gram–second) system, the value is 981cm per second every second. The numbers are approximate (being rounded to a whole number). The value of g is not, strictly speaking, a constant, but varies very slightly from place to place owing to the irregular shape of the Earth. The Earth is not a perfect sphere, but (formally speaking) an oblate spheroid, further misshapen by, for instance, mountain ranges, deep valleys and oceans. These deviations from the true spherical form mean that the Earth's centre is not quite the same distance from the surface at all locations, and cause g to change from place to place.

Isaac Newton

Founded by Galileo, the science of dynamics was enhanced by a succession of innovators and completed by Isaac Newton. Newton's system lasted more than two centuries, before Einstein came along to upset the *status quo*. Newton made contributions – all of them outstandingly brilliant – to several branches of the subject.

Newton always established universal principles by generalising from the contributions of others. For example, he collapsed Kepler's three laws of planetary motion into a single law of universal gravitation. This states that 'every body attracts every other with a force equal to the product of their separate masses divided by the square of the distance between them.' Similarly, in his second law of motion, Newton generalised the concept of force, tying it to acceleration. His work on light was equally impressive. When he separated white light into its component colours, passing it through a glass prism to produce an artificial rainbow, he showed everybody, but especially those with perception, that there was still some relation to be discovered between light and sound, between electromagnetic phenomena and light, and between light and gravity. Einstein devoted many years trying to identify how these forms of energy were related to each other. But even with the best working conditions available to a scholar (no duties, a salary, an assistant, a first-class, not to say magnificent, library, and an unlimited supply of pencils and paper) he had to confess failure at the end of the day.

Johann Doppler

Johann Christian Doppler (1803–53) was Professor of Physics in the University of Vienna. In 1842 he published an article on the coloured light of double stars. The principle discovered by him gave us a method for determining the speed of stars, galaxies and other self-luminous bodies as they move towards or away from the observer. Doppler's principle had probably been noticed for centuries beforehand, though no one had remarked on it.

In the days when a fanfare on the bugle announced the arrival and departure of the stagecoach, the Doppler effect must have been particularly apparent. The ideal way to observe it nowadays is to stand on a railway bridge, preferably in the country, while a train passes underneath. As the train approaches, the driver gives out a loud warning whistle. The observer on the bridge hears the whistle of the approaching train, perhaps sounding the note middle C (that is, air vibrating at the rate of 256 times per second). If the train is travelling at, say, 60 mph (100 km/h) an observer on the bridge is easily aware that the pitch of the note steadily rises to a maximum (say D sharp) until it passes under the bridge. Immediately the whistle is past the observer, the pitch of the note drops back to middle C, or indeed may go even lower. These changes in pitch, which take place quickly, are due to the effects of the train's motion on the sound waves produced by the whistle. The whistle is actually sending out the same note – the same vibrations – throughout. It is the effects of the train, running at speed, on the surrounding air that produces the illusion. The pitch of the mechanical noises made by the train also changes in the same way as does the whistle, the only difference being that the noises aren't musical. The explanation of the illusion is that sound waves that precede the train up to the bridge are 'pushed' together in front of the speeding train and then, as the train passes the observer above, are 'strung out' behind. The wavelengths, and therefore the pitch of the notes heard before and after the bridge, change almost instantly, and so the effect is noticeable.

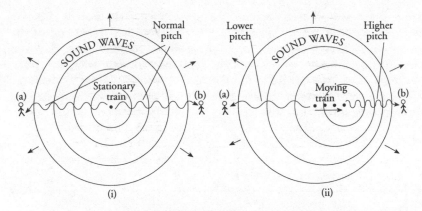

Figure 5   Doppler's principle

The Doppler effect (1842) is, perhaps, the most important principle in the study of how the universe works. The stationary train is heard emitting exactly the same pitch in its whistle from any point around; a moving train (sounding exactly the same note) is heard higher as it approaches, with a sudden drop in pitch when it passes the listener. All radiant energy, including stellar spectra, behaves in this way.

In 1842 Doppler used his law to determine the velocities of double stars relative to each other. The key fact is that light and sound are both wave motions, so that similar rules apply to each. Observing a star through a telescope, it is therefore possible to decide whether the star is moving relative to the Earth (see Figure 6). We can establish which substances are present, as these are represented in the spectrum produced when light from the star is directed through a spectroscope before passing through the telescope.

Basically, a spectroscope consists of a glass prism which splits the light into various colours (and black bars, if the light first passes through a small elongated slit). Using the relative positions of the lines in the spectrum, they reveal the chemical composition of the celestial body by breaking down the light that it emits. It is rather like fingerprinting the star to identify the elements of which it is made.

In the case of fixed stars (those travelling at the same speed as the observer), or by means of a laboratory experiment, we can identify what we might call the 'normal' spectrum. This is when a given star or galaxy, the source of light, is neither receding from the Earth, nor approaching it. However, when a galaxy or star moves towards or away from us at great speed, the spectrum changes. If the Earth observer and the star are moving away from each other, the spectrum colours and the black lines (called Frauenhofer Lines, after their discoverer) also move; they are 'shifted' towards the red end of the spectrum. If the two bodies are approaching each other, the shift is towards the other (blue) end of the spectrum. Although the amounts of these shifts are quite small, they can be measured and then used to calculate the speed of approach, or retreat, of the star or galaxy relative to the Earth. The principle is exactly the same as using a shift in the pitch of a train whistle to calculate the speed of a train relative to the observer.

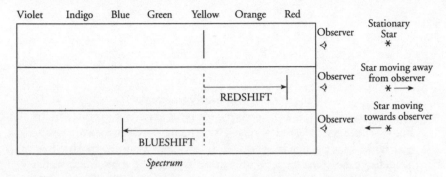

Spectrum

Figure 6   Redshifts and blueshifts in spectra
Rapidly approaching or receding stars or other self-luminous bodies alter in appearance (colour) in a similar fashion to the Doppler shift with sound. Motion towards the observer causes a shift in wavelength so that the blue band in the spectrum is bluer than otherwise. Conversely, motion away from the observer results in the starlight appearing more red.

Information from far-away places

Sitting in one's room in full daylight, contemplating how the world, with its enormous variety of animals and plants and its strange and exotic places, may have come into being is a fairly easy matter. There is no dearth of explanations. Scholars who prefer a rather parasitic (or, more charitably, a 'contemplative') lifestyle like to work out explanations for things in this way, and to make the story sound as trustworthy and complete as possible. But the snag in myth, which revealed itself as time passed, is that revelation is *not* complete, and is certainly not trustworthy. Unfortunately, in the age of myths, there were no credibility tests which could be applied to decide each myth's status as a historical statement; it dealt in certainties, not questions. There was no general agreement on which one legend, out of several hundred, was to be preferred as a record of how things were or how they came to be so.

Stepping outside our room into the evening darkness, we are deprived of our normal resources. It is as though we are in a different universe and have come face to face with the mysteries of creation. To most of us it is an uncharted wilderness of great black expanses of empty space lit up by millions of stars. Witnessing this it is difficult to accept that, for example, it could have all been created in seven days as a home for the human race, let alone that it was forfeited because our first parents succumbed to a carnal urge.

Einstein's great mind could not accommodate the notion that the 'Old One' solved all His important (and not so important) decisions as if by the throw of dice. (He – Einstein – was referring to the principle, enunciated by Bohr and Heisenberg, that it was chance that decided atomic structures and physical events.) He insisted that changes such as an electron jumping from one atomic orbit to another must have been caused by more than someone merely trying to observe its velocity. All physical events, he believed, were the result of a pattern or conjunction of preceding events, including games of chance or decisions made by tossing a coin. Each 'head' or 'tail' was seen as the outcome of a causal process, the only difference being that it was not possible to predict accurately each separate outcome because we didn't have access to all the factors involved to enable us to measure them. However, because there are also laws of probability we could still make an informed guess at heads or tails too.

By contrast, Bohr and Heisenberg believed that we can only predict the statistical outcome of single events. In addition, the act of observation, they said, affects the law of causality. There was no way to discover how one electron rather than another was singled out to make 'the leap'. Einstein continued to insist that 'God does not play dice'. (It has, unfortunately, become an overworked slogan, but it serves to make the point.)

Back in the garden, having spent a few hours watching the skies and keeping some sort of record, we may decide on a change of pace. Now is the time to split our awareness of the celestial bodies from the inchoate feelings of wonder, awe, mystery and other immediate and direct emotions. We need to work on the problem of how to discover, if we can, how everything fits together, to become as familiar with our night environment as with our daytime one. All that is needed, in fact, is an extension of the method for making an accurate map of the territory, or 'estate', inherited from our celebrated ancestors.

Obtaining celestial data

By excluding our spontaneous emotional and personal reactions, we first 'de-centre' the problem and develop an understanding that we live in a kind of sidestreet, off the main track. Far from being at the centre of everything, the Earth is but a small suburban residence. But like suburbs everywhere, it has connections with the metropolis. We are involved in two different jurisdictions. One is the immediate celestial neighbourhood we observe by going out and looking up at the night sky. The other is the same area seen by daylight. Our night environment has been extended and made habitable by modern technology: artificial lighting and modern transport transform the night into a copy of our daytime ambience.

The separation of our lives into day and night is clearly due to the Sun, visible during the day but not at night. It takes a long time to appreciate that the Sun is the centre of our affairs, as it apparently moves across the sky while we seem to be motionless watching it. But it takes even longer to work out that the Sun is still 'out there' in night time, illuminating 'our' heavens even on the darkest night. We can't see anything because the Sun, and its effects, are out of our line of vision. The Earth is facing the 'wrong' way. The Sun does not shut down during 'our' hours of darkness; it works on a 24-hour schedule; when the Earth is turned away, it is we who are in the shadow, who are on the 'dark side', as if in an inner, windowless room during the day.

In compensation, we have a privileged position, as in an enormous open-air theatre. There is a backdrop of twinkling stars which do not move with respect to each other but *en bloc*, and slowly. The fixed stars maintain their positions in the pattern. Moving across this pattern are the planets, with speeds which vary between 10,000 and 100,000 miles per hour (16,000 and 160,000 km/h), but so far away that we can barely detect the motion of 15 degrees per hour. Just like the Sun, they are there during the daytime as well as night; we don't see them, either because we are facing the wrong way, or because our vision is not discriminating enough to single out the lesser lights against the power of the Sun.

There are less than a dozen planets in the solar system. They move in the same direction as the Sun, but all at different speeds. As of today we know, because we can see them through a telescope, that the solar system consists of nine nearby planets in orbit – Earth, Mercury, Venus, Mars, Jupiter, Saturn, Uranus, Neptune and Pluto. Maybe there are others in adjacent galaxies, but we know nothing of them. Some, such as Jupiter, Saturn, Uranus and Neptune, are extremely massive; the others range from Earth size (Earth, Venus, Mars, Mercury) to the quite tiny Pluto. (Pluto is accorded honorary status as a planet mainly for sentimental reasons. It is so small and distant as to be practically invisible; it has one satellite moon, Charon. If we include Pluto among the planets, several similar, small and irregular objects might also be included.)

The massive planets have a number of moons, or satellites: 17 for Saturn, 16 for Jupiter, 15 for Uranus, 8 for Neptune. In the less massive group where we live, Mars has two moons, Earth has one and Pluto has one. Lastly, we have the Sun, which seems to move at an even pace across the sky, appearing at dawn in the east and disappearing in the west at evening. It rotates slowly, but we can't observe this without endangering our eyesight. It is a material body, but not as solid as it looks. The interesting fact about the planets is that each moves on its own select path. This path alters over time (years or even centuries), but very slowly and almost imperceptibly. We can picture the planets travelling at different speeds, as if in a race towards a common objective.

The massive planets have atmospheres largely composed of hydrogen; the four less massive ones all have different atmospheres. Mercury's atmosphere is mainly helium with some hydrogen, while the atmospheres of Venus and Mars are mostly carbon dioxide. Earth's atmosphere is mainly nitrogen mixed with about one-third oxygen – just right for the lung-breathers who have evolved to be fit for it. Pluto's atmosphere seems to be of methane.

Apart from Earth, none of the planets could support any form of life as we know it. In spite of the enormous sums of money still spent on projects to contact 'aliens' in outer space, we are virtually certain of this. There may be vast empires 'out there' to conquer, but there is no positive evidence. The massive daily temperature changes and the absence of water, the lack of any signs of life of even the simplest kind on Mars, which has been studied close to, all point to a total absence of living matter on any other planet in the solar system.

But how do we know all this? Before we turn to an explanation, there are still some relevant facts which should be mentioned about the solar system. The roll-call of the planets in this system, and of their salient features, pinpoints the precise measures we need to define, and supports our judgement about other forms of life.

The Solar System

The Astronomical Unit (AU) is the distance of Earth from the Sun, that is, 93 million miles (150 million kilometres). The planets are spherical, sometimes irregular in shape, as are their satellites.

> (1 km = $\frac{5}{8}$ mile; 1 kg = 2.2 lbs approx)
> 1 million = 10^6 or 1,000,000; Water freezes at 0°C (273°K), boils at 100°C (373°K. –273°C = Absolute Zero, also known as 0°K (for Kelvin). Kelvin units are the same size as Celsius units.)

Earth
Third planet from the Sun, only habitable planet
Orbital period	365.25 days
Mean distance from Sun	1 AU = 150,000,000 km
Diameter	12,760 km
Mass	6.0×10^{25} kg
Density (Water = 1 g/cm^3)	5.5 g/cm^3
Surface temperature	(−60 to 120) °F; (−50 to 50) °C

One satellite, the Moon, sometimes seen in daylight.

Mercury
First planet from the Sun, uninhabitable
Orbital period	88 days
Mean distance from Sun	0.39 AU = 58×10^6 km
Diameter	4,880 km
Mass	0.06 Earths
Density	5.4 g/cm^3
Surface temperature	(−360 to 800) °F; (−220 to 430) °C

Mercury is hard to see because it is so small and close to the Sun. At certain seasons, it is visible just after sunset or before sunrise. Neither it nor the other planets can be seen in daytime because of the glare of the Sun.

Venus
Second planet from the Sun, uninhabitable
Orbital period	226 days
Mean distance from Sun	0.72 AU = 108×10^6 km
Diameter	12,100 km
Mass	0.95 Earths
Density	5.2 g/cm^3
Surface temperature	900 °F; 480 °C

Venus is the brightest object in the sky after the Sun and the Moon. It is often visible for several hours just after or before sunrise (hence it is both the 'evening' and 'morning' star.) It has been visited by spacecraft from both the United States and the former Soviet Union. Much of its surface has been mapped by radar. Viewed through a small telescope, Venus goes through phases similar to those of the Moon. Larger instruments are needed to reveal details because thick clouds cover the entire planet.

Mars
Fourth planet from the Sun, uninhabitable
Orbital period	686 days
Mean distance from Sun	1.52 AU = 228×10^6 km
Diameter	6,800 km
Mass	0.53 Earths

Density	3.9 g/cm^3
Surface temperature	(–271 to 72) °F; (–168 to 22) °C

Mars appears red, even to the naked eye, because of the red iron oxide in the soil. In the Martian spring the surface of the planet changes colour, as the seasonal cosmic winds first blow away the fine reddish dust and then re-cover the darkened surface with it. Mars has two small moons: Phobos is the larger and zips around Mars in $7\frac{1}{2}$ hours; Demos is smaller, and takes 30 hours to make the same journey.

Jupiter

Fifth planet from the Sun, uninhabitable

Orbital period	11 years, 321 days
Mean distance from Sun	5.20 AU = 778 × 10^6 km
Diameter	143,800 km
Mass	317.89 Earths
Density	1.3 g/cm^3
Surface temperature	–200 °F; –130 °C

Jupiter is the largest planet in the solar system. In composition it is more like a star than a planet. Bright belts, changing cloud structures, four main moons (each larger than the Earth's moon) and the Great Red Spot are all easily visible through even a small telescope.

Saturn

Sixth planet from the Sun, uninhabitable

Orbital period	29 years, 168 days
Mean distance from Sun	9.54 AU = 1,427 km
Diameter	120,000 km
Mass	95.15 Earths
Density	0.7 g/cm^3
Surface temperature	–274 °F; –170 °C

Saturn is perhaps the best-known astronomical object in the skies because its magnificent rings make it so striking in photographs. In close-up its beauty is breathtaking. Magnification reveals many details of the delicate ring structures as well as bringing out the coloured bands in the outer layers. The rings are composed of billions of particles of water ice and methane ice. These icicles are solid objects ranging from a few centimetres in diameter to several metres. There are three major ring systems separated by divisions known as the Cassini (outermost) and Encke (innermost) divisions.

Uranus

Seventh planet from the Sun, uninhabitable

Orbital period	84.01 years
Mean distance from Sun	19.2 AU = 2,870 × 10^6 km
Diameter	52,300 km

Mass	14.54 Earths
Density	1.2 g/cm^3
Surface temperature	below −250 °F; −150 °C

Until spaceship *Voyager 2*, little was known about this distant planet. Although Uranus is visible even through small telescopes, little is revealed to Earth observers apart from its pale blue colour. The planet is now known to have a system of at least nine rings.

Neptune

Eighth planet from the Sun, uninhabitable

Orbital period	163.79 years
Mean distance from Sun	30.06 AU = 4,497 × 10^6 km
Diameter	49,500 km
Mass	17.23 Earths
Density	1.7 g/cm^3
Surface temperature	below −250 °F; −150 °C

Although similar in size and composition to Uranus, the eighth planet is so far from Earth that ground-based observations reveal very few details. Data collected by *Voyager 2* confirm the existence of a ring system. Neptune's biggest moon, Triton, may be observed with some larger telescopes.

Pluto

Ninth planet from the Sun, uninhabitable

Orbital period	248 years
Mean distance from the Sun	5,900 × 10^6 km
Diameter	about 5,500 km
Mass	approx. 0.03 − 0.1 Earths
Density	uncertain, very low
Surface temperature	uncertain

The status of Pluto is uncertain; it differs in many ways from the other planets. It is very small and very remote and probably has a different origin. It may be a large asteroid or even an escaped satellite (moon), thrown off by Neptune. It is very irregular in shape.

Measuring distance

To rephrase our earlier question, how is it possible to have such clear-cut, precise and numerical data and details of conditions on astronomical bodies many millions of kilometres away? Obviously, before we can make up our minds about the authenticity of the reported facts we first need to know how they were obtained. Otherwise we risk leaving ourselves open to the same charges levelled against Plato. We need to know how measurements of location, distance, position, velocity, weight, chemical composition and period in orbit are determined.

Triangulation

The problem of measuring astronomical distances to make an accurate map of the universe is no different in principle, only in the scale of the operation, from the problems that confronted early map makers. In 1804, William Smith set out to produce the first geological map of England. He used a method called triangulation, which is now standard for such surveys.

First we choose three points, A, B and C, and at B and C we ask an assistant to erect vertical survey poles (see Figure 7(a)). We are standing on the third point, A. Now, using a theodolite (an instrument that measures angles between objects at a distance), we check off the angles formed by the imaginary lines between us and the poles at B and C. The distances between the poles can be measured either by pacing, or by means of a lightweight chain carried for the purpose. We then record these results by making a sketch in our notebook. We have already moved away from our starting point, that is, from A to B. So we plant a survey pole at a new point D to establish a triangle BCD. We then measure (again by pacing, or by chain) the distances BC and DC. We proceed in this way, building up our results and recording them as we go, in a notebook (or on the developing map of the geology or archaeology or other special features). In this way, we walk all over the project area, blocking in our findings routinely at the end of the day. This is how it was done 200 years ago. Today we can short-circuit the procedure by purchasing sheets from of blank outline maps (for example from the

(a) *Geographical*

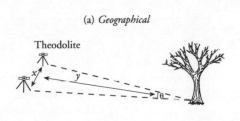

(b) *Celestial*

Figure 7 The method of triangulation

Geological Survey in Britain), and record our work (geological or other) as we go along.

Parallax

To survey the distances of far-off celestial bodies, we use the comparable method of parallax. The survey will be more precise if simultaneous measurements are made by two people, one at each end of the Earth; or better still, if the measurements are made at a six-months' interval from points at each end of the Earth's orbit (see Figure 7(b)). Surveying progressively larger and larger areas allows us build up a map of the surrounding celestial 'territory'. This method was known and used by Oriental scientists and by the ancient Greeks. The scale of measurement can be extended by observing suitable planets on special occasions – for example, the transit of Venus (its passing across the Sun's face), or the distances of the asteroid Eros in its path across the Sun (first observed in 1930). Using two widely separated points on Earth or on Earth's orbit as baselines, accurate measurements can be made.

Instead of parallax, other methods can be used. For example, calculating the distances of particular stars or galaxies allows these objects to serve as 'markers' or 'beacons' for singling out other stars of the same magnitude (luminosity and distance). The fact that the magnitude is the same means that these stars are at the same distance from Earth as our markers.

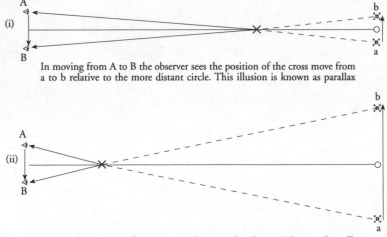

(i)

In moving from A to B the observer sees the position of the cross move from a to b relative to the more distant circle. This illusion is known as parallax

(ii)

The shift is greater if the cross is closer to the observer. The parallax effect is seen in some nearby stars as the Earth moves around the Sun. By measuring the amount that these stars shift relative to more distant background stars, the astronomers can calculate the distance to them

Figure 8   The use of parallax to measure distance

Lastly, since about 1945 we can measure distances accurately by bouncing a signal off a celestial body and measuring the time it takes for the signal to make the return journey. A caesium clock can measure these time differences to an incredible degree of precision. The measurements made by other methods can be verified by comparing them with those made by radar. Using two different methods to obtain our celestial measurements gives us more confidence in our results. For example, by traditional methods of measurement, the astronomical unit was taken for a long time to be very close to 93 million miles. This distance between Earth and Sun has now been verified by radar to be exactly 92,975,699 miles, and some decimals.

Parallax is a phenomenon which has become so natural to us in our everyday lives that we consistently use it, and accept it, without conscious thought. It ensures that we perceive things in depth, as occupying three-dimensional space. It can be defined as the apparent change in the position of an object due to a real change in the point from which it is observed. Its effects can be seen by focusing on a distant view, then closing and opening, first the right eye and then the left. Although our head remains in the same place, our two eyes are about 4 inches (10cm) apart, so the effect is as though we had moved our head four inches. Objects in our one-eyed line of vision seem to move across this line, in the opposite direction to the movement of the objects further away. This is shown in Figure 8. The effect can be tested by moving around the room while looking out the window, closing and opening first one eye, then the other.

Measuring orbital velocity

Galileo's laws of motion, based on experiments in the laboratory, not on leaning towers, were revamped by Sir Isaac Newton. They were incorporated into Newton's system as basic axioms or extended definitions, similar to propositions in geometry. Together with his law of universal gravity, they provided a calculus (a set of procedures) which could predict movements of bodies on the ground or in the solar system, or even far-off galaxies. Kepler's laws of planetary motion underlie the Newtonian system and are also used quite directly in calculations. Only bodies moving at or near the speed of light (186,000 miles or 300,000 km per second) are exceptional, and have to be dealt with in the special context of Einsteinian relativity.

Newton, and others who followed Copernicus and Galileo, was involved in solving classic two-body and three-body problems. Heavenly bodies, and earthly ones too, behave as though all their mass is concentrated at a single point, called the centre of gravity. In the case of spherical heavenly bodies such as planets, the gravity centre is the centre of the sphere. Newton also demonstrated that the orbits of planetary-type moving bodies always took the form of a conic section: namely, a circle or an ellipse if their velocities were too small to allow one body to escape; a parabola or hyperbola (open and smooth curves) for meteors, asteroids or even satellites, when the speed of one of the bodies for some reason exceeds the escape velocity.

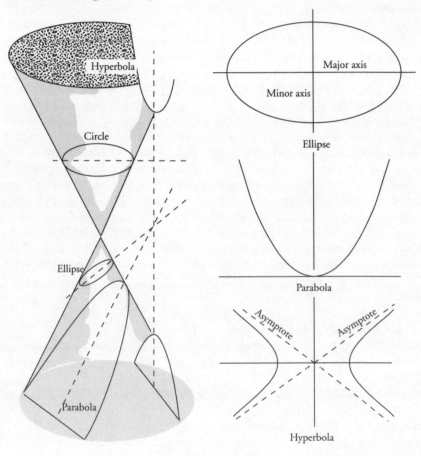

Figure 9   Conic sections
Sections through a cone at different levels give rise to a parabola, ellipse or hyperbola depending where the cut is made. This was important in showing that planetary orbits (and the heavens) worked on geometrical principles.

The chemical composition of the universe
In 1666, Isaac Newton discovered that, when he passed sunlight through a glass prism, the light appeared on the wall of his study as a beam of

merging colours – red, orange, yellow, green, blue, indigo, purple –
indeed all the colours of the rainbow. It was as if a ghost had appeared
through the wall, so he gave this unique set of colours the Latin name for
an apparition, which is spectrum. The rainbow is a spectrum.

Later, scientists discovered that, if you burned salts of various metals,
such as table salt, by dropping a tiny spoonful into a Bunsen burner, you
would see the whole flame as a single colour – yellow for sodium, red for
calcium, and so on. This was an epochal discovery. Earlier, to analyse an
unknown substance, a chemist (that is, not a pharmacist but someone
trained in chemical analysis) might spend hours testing for each element
in turn. It was a very elaborate procedure, checking for each element in a
solution of the salt, first for the metal and then, using another set of
procedures, for the acid radical. After 1859, when Bunsen popularised his
famous burner as a standard piece of laboratory equipment, a tiny
amount of the solid substance could be dropped into the flame, and the
colour of the flame would identify the metal immediately.

The next crucial observations were made in Britain by Wollaston, and
then in Germany by Frauenhofer who was a contemporary of Bunsen
and his associate Kirchhoff. Whereas Newton had simply diverted his
sunbeam through a ragged hole in his curtain, leading the sunbeam to
pass through the prism, Frauenhofer discovered that if you used a small
slit (a diffraction grating) the spectrum of sunlight still appeared, but the
several colours were now crossed by a great number of parallel black lines
(similar to the bar-codes used nowadays to identify supermarket goods).
Frauenhofer also discovered that the same lines, or groups of lines, were
produced when the material was burned in a Bunsen flame, and the light
was first passed through a mist of the same incandescent material before
passing through the prism. This was a laboratory simulation of what can
be seen happening in the Sun when it is viewed through a prism. Many
elements, such as hydrogen, are consumed by the Sun. The sunlight
produced by internal nuclear reactions has to pass through a wall of
burning gas (partly hydrogen). This combination of events acts to generate the black bar-code line in exactly the location known to be the
wavelength of sodium.

Frauenhofer's experiment was crucially important. It gave rise to a new
science – solar spectroscopy. Suddenly, large numbers of scientists
became life-long addicts devoted to the task of identifying the patterns
of hundreds of lines in the spectrum of light from the Sun and other stars.
Thanks to their efforts, we can tell exactly which elements are present or
absent in the Sun and in other self-luminous bodies.

At Harvard and other institutions the new science created a perfect
frenzy of activity. One distinguished astronomer set up what his friends
referred to as 'the Harvard harem'. This consisted at one time of about

two dozen or so female secretaries who were paid by the hour to classify thousands upon thousands of photographs of solar spectra. As one result, several women became well-known astronomers in what was at the time considered to be a strictly male preserve.

The element helium was first discovered on the Sun (and was named after Helius, Greek god of the Sun) by Sir Norman Lockyer, long before Sir William Ramsay at Glasgow University isolated it as a rare gas in the Earth's atmosphere. Although rare on Earth, it rivals hydrogen as one of the most important substances in the universe.

Chapter 4

Continuity in Nature

It is all too easy to derive endless strings of interesting-looking but untrue or irrelevant formulae instead of checking the validity of the original premises.

John Ziman

Classification: the basic method of science

Clearly, from the previous chapter, one way to build a solid foundation for science is to handle the raw materials. This involves direct contact with data rather than an uncertain, speculative affair with theory. Armed with a large selection of materials, we can proceed to the next stage. This consists of classifying the data, using the minimum number of categories necessary to yield a reasonable system. Examining raw data can help us discover principles which might explain the origin of our cases and a possible time-scale of development. If we are successful, these procedures can explain the variety, and account for the similarities and differences, of the raw materials chosen.

Here, we consider the origins of the planets, and especially of the Earth. Unravelling and dating the geological sequence of Earth provides an excellent model for developing the science of the universe. The following section leads quite naturally to the key question, how can we account for the origin of cosmic systems?

How old is the Earth?

The French Revolution at the end of the eighteenth century was a new intellectual beginning. Many attitudes and ideas previously forbidden by Church or state, or neglected because of state or private censorship of public opinion, were revived. Views which had been suppressed for centuries became viable subjects of conversation in drawing-rooms, and more importantly, of investigations in laboratories. In spite of local excesses, such as the execution of Lavoisier for his alleged involvement in 'farming' taxes for the monarchy, science was fostered and given favoured treatment, especially under Napoleon. He, at least, understood

51

that rulers who gave moral support to scientists, and to scholarship in general, stood a better chance of surviving in the modern world than did the vacuous aristocracy.

There are stories that Napoleon, as an army engineer, liked to throw his scientific weight about. One of the most quoted ones is how he, having read Laplace's book on celestial mechanics, challenged the author that he could find no mention of God there. No, said Laplace, he had found that he didn't need 'the God hypothesis'.

This was a time when several sciences, including geology, were being revived. Earlier, they had caused disquiet because certain remarks were identified for political purposes as anti-religious. Revealed texts may be necessary to establish religious credibility, but for testing scientific discoveries they are irrelevant. Certainly they contain none of the fine detail which is usually the essence of a scientific explanation.

Unfortunately the fundamentalist view of the Scriptures does not care for any alternative view of Creation, or scepticism about events such as the Sun 'standing still' at God's command, or stars appearing in the skies (however briefly) to mark some religiously significant event.

Geologists have taken issue with fundamentalists (who claim that every word in the Bible is divinely inspired and therefore true), because the first chapter of Genesis is unacceptable as a literal account of the beginning of the world. The sequence of events described there is quite different from the one established by the comparative study of the fossils of animals, plants and fish. From the evidence of fossil remains excavated from rock strata, and capable of being assigned precise dates in a time sequence, the time-scale of Genesis was absurd; the Earth was clearly very much older than theologians' estimates (such as that of Archbishop Ussher of Armagh in 1650, that if you add the generations of family trees given in the Bible, you will find that the date of the Creation was 4004 BC).

The actual sequence discovered in the rocks was much richer and much more interesting than these narrow, anti-historical accounts. In the 1860s Thomas Henry Huxley argued the case with those of his contemporaries foolish enough to claim supernatural guidance about fossil sequences. Since then, what Plato might have called 'proper' zealots have been wise enough to steer clear of such matters as having no relevance to the truths of their religion.

The law of superposition

In many geological phenomena, such as the deposit of sediments by rivers, the Earth behaves as a massive hourglass measuring the passage of time. Rivers wash away sand, transport it as silt, and then deposit it in still waters as thick layers. These materials are converted to solid sandstone or mudstone by the weight of overlying materials. Enormously thick beds of rocks, such as the limestone cliffs of Dover, or the sandstone strata in the Grand Canyon of Arizona, testify to the lapse of very long periods of time.

The duration of a sedimentary process can easily be measured by assessing the rate of deposition of the same materials in any modern river. This principle, called 'uniformitarianism' was enunciated by the Scot James Hutton in 1785. Its basis is that geological processes have, throughout history, been the same as those which operate in the here and now. Also important is the law of superposition: that, in any undisturbed sequence, the oldest rocks, being deposited first, must be at the bottom, and the youngest at the top. Taken together, these principles give us a method for correlating the ages of different out-crops on the ground.

Hutton's rules, and the details of how they operate, were a new charter for geology. They are an exact analogy of Galileo's principle that the laws of physics are universally valid, in the Heavens as on Earth, and Herschel's principle that farther-off celestial bodies are older. As these gave new meaning to astronomy, so Hutton's views created a new theory of the Earth. Identical types of rocks, fossils and rock sequences enable the geologist to reconstruct the course of events over regions and across continents. From evidence directly under his or her feet, the geologist can determine the whole national time sequence of events in (say) Britain, China or Africa, and proceed to do the same on a world scale.

Since the pioneering geological maps of Britain were made in the early 19th century, most national governments have followed suit. There are geological maps of the Russian land mass, Germany, France, Holland and most other places. Remarkably, although only parts of the complete sequence can be found in any given area, these pieces fit like parts of a jigsaw into the same broad picture. The British succession of geological systems is typical of systems found elsewhere. Not only was Britain first in the field in modern times, but it is also fortunate in having almost have every kind of rock, from the oldest rocks in the northwest Highlands of Scotland through to some of the most recent deposits of the great Ice Ages. Looking around open country, with a sophisticated geological eye, immediately brings us into contact with a past which extends to the beginning of the time on Earth. Much the same is true of the universe.

Radioactive dating

The most precise clock records are provided by radioactivity. This was studied in 1898, by Maria Sklodowska Curie and her husband Pierre. One substance used for dating naturally occurring materials is radium. This is because it disintegrates at a constant speed and is totally unaffected by temperature, pressure or anything else. It has a 'half-life' of 1,620 years. In other words, its atoms give off particles or rays at a constant, unvarying speed, as the radium gradually changes into lead in the process. Radium is one element in the chain of disintegration from uranium 238 to lead. The half-life of uranium 238 is 4.5 billion years. The rate of disintegration is precise and constant. But, as it acts on fewer and fewer atoms over time, the number of atoms which break down is also reduced. This leads to an infinite regress, so the *total* change to lead would take an infinity of time. To provide the rate value, scientists therefore quote the 'half-life' – the time it takes half the original amount to break down.

In the chain each atom has a certain average probability of breaking down at any given moment. When an atom breaks down, the quantity of radium, for example, in the sample is reduced by one atom, which is replaced by one atom of lead. (Incidentally, there is no way we can predict which atom will disintegrate, or when. This is a second example of the principle of indeterminism. The first was the failure of the possibility of prediction in games of chance which was studied about three hundred years ago.) In exactly 1,620 years, half the number of radium atoms will have disappeared, having been replaced by atoms of lead. Both the time series for the decrease of radium and for the increase of lead are *infinite* mathematical series; the one goes on getting bigger and bigger, as the other becomes smaller and smaller. This fact, that the increase in lead atoms is paralleled by an equal decrease in the number of radium atoms, means that, regardless of the age of the radium, the weight of lead divided by the weight of radium yields the age of the material in which the two metals are found in close association.

As well as radium, other radioactive atoms are also used for dating. If, for example, the sample to be dated is organic (made of wood or other organic material), radioactive carbon 14 is used.

Radioactive dating gives an exact or absolute value for the age of the rocks, unlike the law of superposition which only tells us in whether rocks are 'younger' or 'older' than each other. However the two estimates can be used together, each confirming the other. The results of these estimates and calculations around the world for over two centuries are summarised in the table below.

The geological Time-scale and the age of the Earth
Youngest rocks are at the top

Era	Period	Duration	Time estimate
Cenozoic	Quaternary	7 million	present day to
	Tertiary	years	7 million years ago
Mesozoic	Cretaceous	160 million	167 million years
	Jurassic	years	ago
	Triassic		

Palaeozoic	Permian	300 million	467 million
	Carboniferous	years	years ago
	Devonian		
	Silurian		
	Ordovician		
	Cambrian		
PreCambrian	PreCambrian	4,030 million	5,000 million
	Division	years	years ago (in
	(local series		round figures)
	names differ		
	by country)		

Oldest Rocks are at the base

There has been much argument about the age of the Earth. In the 1890s Lord Kelvin reckoned that the Earth was a mere 40 to 100 million years old, basing his calculation on an estimate of heat loss by the Earth from its beginning. Since then the estimates have grown and grown. Kelvin's calculation was made before the heat donated by radioactivity was recognised. As new generations of geologists became interested in the question and less afraid of big numbers, the time scale expanded. It is now established beyond doubt that the Earth came into existence many millions of years ago (in round figures probably five thousand million years). It is still hard to deal with numbers as vast as these.

The origin of the solar system

The planets in the solar system travel in ellipses which are spaced out almost in a single dimension, rather like the grooves on a gramophone record. There is a slight 'wobble', as the orbits are just off parallel. In other words, the planets don't 'wander' about the sky any more than the fixed stars do. Their elliptical orbits have a single common focus (the Sun), and they move almost (but not quite) in a single plane. These two facts are indeed extraordinary. They can be explained only by a general principle which ties the physical properties of the planets in with the operation of some simple, physical law. The facts that solid objects, such as meteorites, appear from outer space, and that they are similar in appearance and composition to rocks and minerals found on Earth, also strongly suggest a common origin for all the materials and mass of the solar system.

The Earth itself has a remarkable amount of what can be identified as once-molten rock. These rocks are found especially in mountainous country as igneous and metamorphic rocks. They can be recognised with ease because they differ in appearance and structure from sedimentary rocks, such as limestone and sandstone, and from unaltered volcanic rocks, which are more uniform in their structure and appearance.

In 1755 the German philosopher Immanuel Kant (1724–1804) put forward what he called 'a theory of the heavens'. (The theory was developed independently by the eminent French mathematician and astronomer Pierre de Laplace, who was ennobled by Napoleon as a marquis.) Kant tried to explain facts such as the following: all the planetary orbits are almost circular and almost in the same plane; the planets move around the Sun in the same direction as the Sun itself in its rotary motion; and the satellites move around the planets, also in the same plane, in almost circular orbits. Since Kant's day we have added the following relevant facts: satellites have a total mass just about one-thousandth that of the planets, and the planets together weigh about one-thousandth the weight of the Sun; the 'Earth-size' planets (Mercury, Venus, Earth and Mars) are all small, dense and relatively close to the Sun, whereas the giant planets (Jupiter, Saturn, Uranus and Neptune) are all large, of low density and relatively far from the Sun; planetary distances from the Sun are largely in agreement with the Titius–Bode law. This states that, if the Earth's distance from the Sun is taken as the Astronomical Unit – 93 million miles (149, 500,000 km) – then planetary distances can be expressed by the series

0, 0.3, 0.6, 1.2, 2.4, 4.8, 9.6, 19.2 . . .

The Sun rotates very slowly, its angular momentum being only 0.5 per cent of that of the total planetary system.

The Kant–Laplace explanation is called the 'nebular hypothesis'. (It was actually about galaxies not nebulae, but very little was then known about either.) In effect, it states that the solar system came into being as a vast accumulation of magma (volcanic molten rock) in space, spinning round and round, all the time cooling and therefore contracting. As we know from watching an ice skater performing the figure known as the 'tight spin', a spinning body can control speed by reducing or increasing the body's extension at right angles to the direction of spin. Thus, holding the arms tightly against the body, or reaching high while spinning, increases the skater's speed, while extending the arms slows the skater down. The principle is known as the law of conservation of angular momentum. Examples can be found in all sorts of activities – high-diving, ballet dancing, flying-trapeze work – indeed wherever a spinning body is involved. Using this principle, in 1982 the trapeze artist Miguel Vasquez performed the first-ever quadruple somersault. It was estimated that his flying body, when caught by his team-mate (the 'catcher'), was travelling at over 70 miles per hour (113 km/h). Dangerous work!

Returning to our cooling and rotating molten magma: as it cools it and contracts, so it spins faster and faster. However, unlike that daring young man Miguel Vasquez, this magma is not a solid but a semi-liquid body, and at a certain speed portions will split off and be ejected at considerable speed, some fragments travelling farther than others. Once ejected, they continue to spin and cool down until they solidify. As these masses are only semi-solid, they will assume a spherical shape due to the gravity. Some, such as the Earth, will continue to rotate about their own axis so that, when solid, they will form the sort of planet we know. The force of gravity, combined with the initial escape velocity, will cause the semi-solid, and then solid, masses to continue to revolve in circular or slightly elliptical orbits. The shape of the orbit is determined by interplanetary and galactic forces.

Unfortunately the Kant–Laplace theory leaves many problems unsolved and is now no longer accepted, indeed current scientific opinion is totally against it. However it does serve as an illustration of the kind of hypothesis that is worthwhile. There is no doubt that the stars and planets, including the Earth, did evolve in some such fashion. The theory thus provides a starting point for thinking about cosmic origins.

Sky objects – the raw materials of cosmology

Before the invention of the telescope, the only optical aid available to astronomers was the quadrant (a navigational instrument for measuring the angle of heavenly bodies against the horizon – an ancestor of the sextant which has invented in 1730). Not surprisingly, for centuries astronomy was confined almost entirely to the study of the solar system – that is, the planets and a limited number of very bright stars. The telescope was invented and reinvented many times; we don't know who was the first. The most important telescope, however, was Hans Lippersley's invention in 1609, because he started making them for sale. Galileo heard of this new device from a friend and had soon made his own and then turned it on the heavens.

However, to start with, the telescope was more a gadget or a symbol than a stimulus for change, and the full-blown revolution in observational methods did not come till later. People (especially the learned) had to be persuaded, over a very long time before they would, with great caution and unease, even handle this sorcerer's device.

As well as disposing of the Ptolemaic view (see page 25), the telescope did eventually produce a mighty revolution in observational astronomy. This was carried through almost single-handedly, nearly two centuries after the first telescopic observation of the Heavens, by William Herschel and his sister Caroline. Herschel (1738–1822) was a musician who, till the age of 19, played in a military band in Hanover. When the French invaded Hanover during the Seven Years' War, William and Caroline escaped to England (where another emigré Hanoverian, George III, was king). Nine years later William was appointed organist at a respectable church (the Octagon Chapel) in Bath. There he happened to read a book about the Pythagorean theory of music, and discovered that its author (a mathematician and astronomer) had also written a book on how to make large telescopes. By coincidence, one of Herschel's neighbours (who had probably lent him the book) had a complete set of tools and materials for building telescopes, and had just decided to give up this hobby. So William, who now had plenty of spare time, and the tools for the job, started to make large telescopes of his own. He then used them to make observations, while Caroline made notes and drew meticulous star-maps.

Herschel was not particularly interested in the solar system, so he turned his new and very large telescope on other galaxies and far-off objects. This symbolic turning of his back on his own cosmic neighbourhood, in favour of the cosmos, launched a new era in astronomy. Until the Herschels proved otherwise, most contemporary astronomers had thought that the solar system and its planets were all very well, but were really for children and beginners.

But before quitting the solar system, Herschel marked his respect by discovering the planet Uranus, on 13 March 1781. It was the first new planet to be discovered since antiquity. He showed his respect in another quarter by naming it *Georgium Sidum* (George's Star), after George III, for which gesture he was awarded a state pension. (His duties, in return, were to visit the palace occasionally and let the royal children look through his telescopes.) Financial security meant that he could give up music as a profession and devote his life to the study of the cosmos.

William Herschel's son John also became an astronomer of high repute. As a student at Cambridge, he was a close friend of the mathematician Charles Babbage. In fact, they were working together on logarithms when Babbage conceived the idea of a computing engine. This provided astronomers with yet another instrument and ushered in the modern age of computers.

Galaxies

Herschel the elder classified the galaxies and speculated about their evolutionary history. In fact Charles Darwin's grandfather, Erasmus Darwin (1731–1802), had already written about the theory of evolution in organic nature, so the notion had been in the air for some time. Hubble's classification is illustrated opposite, and has been extended to other sky objects. It underlies some basic questions with which twentieth-century cosmology has been faced, and is likely to face further in the next century.

Galaxies are now grouped according to shape, as spiral, elliptical, or irregular. Spiral galaxies are very common and can be either a normal spiral (S) or a flat spiral with no arms (S0) or a barred spiral (SB) crossed with parallel lines of fine dust which look like prison bars. The spiral galaxies are further sub-divided into a, b and c types according to the open shape and the sharpness of the spiral. Elliptical galaxies are classified according to how far they diverge from a circular shape. The system goes from E0 (circular or almost circular) to E7 (an ellipse with the major axis three times as long as the minor axis). E1 to E6 are ellipses whose shape lies between these limits. Irregular galaxies are those that are neither spiral nor elliptical.

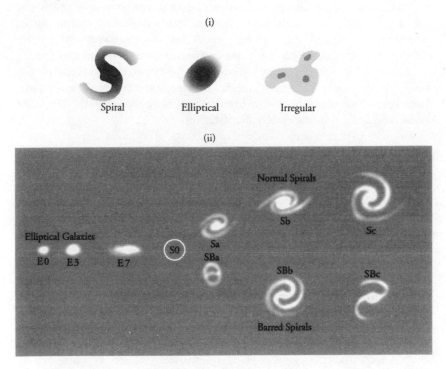

Figure 10   Galaxy types (i) and Hubble's classification ('tuning fork' diagram) (ii)

The galaxy Centaurus A
Object type: Spiral galaxy (S0 type)
Distance from Earth: 14,000,000 light years

Centaurus A is a peculiar and energetic galaxy which has long been the subject of controversy. It is a luminous sphere crossed by a prominent but irregular black band. It has undergone many changes of classification over time but seems to be a huge elliptical S0 (almost circular), receding at between 220 and 320 miles per second (350–510 km/s). It was estimated to be 6.8 million light years from Earth but this estimate was increased first to 15 then to 20, and even 25 million light years. It is as brilliant as 20 billion suns and is one of the strongest sources of radio waves in the universe. It was among the first radio sources to be discovered, when the first dishes to receive radio waves were built at Jodrell Bank, near Manchester, in about 1950, Radio astronomers have since learned that Centaurus A is a double radio source. The galaxy, whose centre is visible, throws out vast amounts of materials in opposite directions at nearly the speed of light. This material is slowed down when it collides with interstellar dust and gas. It then gives out radio waves, which are believed to be produced by these materials interacting with galactic magnetic fields. Gigantic explosions take place in the nucleus of the star, whose whole behaviour seems, so to speak, 'pathological'. Centaurus A is not alone in this; other systems, such as galaxies M82, M87 and Cyngus A, require similar clarification.

Andromeda galaxy in the constellation Andromeda
Object type: Spiral galaxy (Sb type)
Distance from Earth: 2,500,000 light years
Distance across galaxy: 160,000 light years

The Great Galaxy in Andromeda is the largest of the Local Group, which includes the Milky Way, where Earth is located. (The word 'galaxy' is from the Greek meaning 'milk'.) The Andromeda Galaxy is the only one which can be seen by the naked eye, appearing as a small, elongated fuzzy patch of light. It has been known since AD 905 and is the nearest of all spiral galaxies. Containing over 300 billion suns, Andromeda is far bigger than the Milky Way. In our line of vision, it is almost edge-on and tilted at an angle of 13° and is the most distant sky object visible to the unaided eye. In fact, it can be seen only from sites away from the glow of city lights and when the moon is below the horizon. With a telescope, the two satellite galaxies NGC205 and M32 can also be seen in the same field.

There have been a number of conflicting estimates of distance from Earth because of the relatively chancy (cf. page 46) 'beacon' method of estimating distance. But the value given above seems acceptable. In weight the star is equal to about 400 times the mass of the Sun. It belongs with the most luminous galaxies in the universe, being equal in luminosity to 11 billion suns. From photographs, its diameter has been estimated to be 110,000 light years, that is, about 1.8 million million million miles. (One light year = 5.88 million million million miles or 9.465×10^{12} km.) Except by cosmic standards, this is pretty far from Earth.

Nebulas

The different terms used over the years to describe nebulas and galaxies can be confusing. Galaxies are island universes which harbour many stars. Each star is itself a galaxy, that is, an organised system, perhaps something like our solar system. Galaxies can be very large and consist of massive volumes of gas and other thinly spread materials. But the main mass of a galaxy consists of solid bodies – stars and clusters of stars. Nebulas differ from galaxies in that, although their make-up is the same, the proportions are reversed, so that in nebulas the gaseous, dusty, fine-grained solid materials predominate.

Nebulas are therefore mostly, but not invariably, composed of wisps of vapour, dust clouds and other 'nebulous' matter – a quantity of gas with perhaps a star, or star clusters, or a number of separate stars embedded in it. Some nebulas consist of dark clouds of fine matter. These are lit by massive radioactive and other reactions. They can be either the remains of a supernova (an intensely bright star which may have exploded and is visible for hours or days before dying down), or an emission nebula, which sprays out massive clouds of material in several directions, or a planetary nebula, made up of a solid body or bodies, in a thin gruel of wispy or even porridge-thick materials. Some of the better-known nebulas are listed below.

The Great Nebula in Orion

Object type: Emission planetary nebula
Distance from Earth: 1,600 light years
Distance across nebula: 16 light years

The MU2 Nebula is the middle 'star' in the sword of Orion the Hunter. It can be seen without optical aids as a hazy star, but through even a small telescope it is a magnificent sight. M42 is the brightest nebula visible from Earth and is clearly the most studied of its kind. It was discovered by Christiaan Huygens in 1656, and Herschel recorded his observations about it on the very first page of his first notebook. At a distance of 1,600 light years, it is relatively close to Earth. Many new stars are forming in the interior of the cloud and several proto-solar systems have been identified by infrared photography. Six-inch and larger telescopes reveal a cluster of four stars at the heart of the nebula. These are called the Trapezium, since they can be seen as the corners of an imaginary four-sided figure with two sides parallel. These stars are among the youngest yet discovered.

The Crab Nebula in Taurus

Object type: Supernova remnant
Distance from Earth: 6500 light years
Distance across nebula: 7 light years

The Crab Nebula is the remains of a supernova that exploded in July 1054. The explosion was recorded by Chinese astronomers. According to the records, the supernova was visible for several weeks in broad daylight. The star left over was the first pulsar to be identified 900 years later by radio astronomers. It is now known to be a neutron star which revolves on its axis once every 33-thousandths of a second. The Crab Nebula is brighter than 75,000 suns.

The Cone Nebula in Monoceros
Object type: Dark cloud nebula
Distance from Earth: 4,500 light years

Similar to the Horsehead Nebula, the Cone Nebula is another example of a dark dust cloud silhouetted against an emission nebula. The presence of hot young stars indicates that the emission nebula is surely the birthplace of these stars and supports the view that stars are born from collisions of very hot gases (such as hydrogen and helium) with cosmic dust clouds.

NGC6514 The Trifid Nebula in Sagittarius
Object type: Emission nebula
Distance from Earth: 3,200 light years
Distance across nebula: 12 light years

The Trifid Nebula is a glowing pink cloud of interstellar gas lit by a relatively new star (7 million years old). Three dark radial dust lanes divide the cloud, hence its name. The light coming from the hydrogen gas is associated with the red hydrogen–alpha line in the spectrograph.

Associated with the Trifid Nebula is a blue nebular area surrounding the star HD 164514. Unlike the star within the Trifid, this star is not hot enough to excite the hydrogen gas, so all we see is starlight reflected from dust particles and cold gas. These particles selectively scatter blue light, in the same way as sunlight is scattered when it passes through the fine cosmic dust surrounding the Earth to produce the blue of the sky as seen from Earth by day.

Star clusters
Clusters can assume a number of forms but make up an easily recognisable group of several stars fairly close together. The groups may be open or globular (closed), with anything from two to several million members.

The Globular Cluster in Hercules
Object type: Globular star cluster
Distance from Earth: 25,000 light years

The Great Globular Cluster in the constellation Hercules is a spectacular group of hundreds of thousands of older stars which seem to circulate around the Milky Way. M13 is the brightest globular cluster to be seen in the Northern hemisphere.

Jewel Box in Crux
Object type: Open star cluster
Distance from Earth: 7,600 light years

This star cluster is so-named because of the vivid contrast between the star kappa Crucis and the other stars in the cluster. The kappa star is bright red and is surrounded by blue and white stars of different shades and intensities. The whole effect is like a collection of precious jewels, hence the name.

The Pleiades in Taurus
Object type: Open star cluster
Distance from Earth: 410 light years
Distance across cluster: 5 light years

The Pleiades, also known as the Seven Sisters, is one of the finest examples of a young star cluster. It can be easily seen with the unaided eye. Ancient astronomers reported seeing seven stars ('the seven sisters') in this area with the naked eye, but today only six can be seen without an optical aid. Long-exposure photography shows much nebulosity (gas clouds) in the region. It is believed that it was from these gas clouds that the Pleiades stars were originally formed. The cluster is very young, in fact less than 100 million years old. Nearby is an older cluster, the Hyades, of which Aldebaran (the Alpha star in Taurus, and one of the brightest stars in the sky) is a member.

Star names and the history of cosmology
We can estimate from catalogues the relative contributions to obervational astronomy made by the chief nations working in this field. Inevitably the selection may reflect some bias, but the comparisons will serve an as approximation to the proportional contributions.

The sources of our knowledge can be identified from the names given to sky objects. Many date from long ago and originated in the ancient civilisations. We can tell which ones these are from the linguistic origins or the mythological source of the names. An approximate and limited sampling reveals the following distribution:

Number of star names by language/civilization					
Arabic	*Greek*	*Chinese*	*Biblical*	*Egyptian*	*Roman*
1,060	460	1,460*	effectively, very few original names either in literature or other sources		
	* to year 600 BC		no star catalogues exist		

Burnham's *Celestial Observer* lists nearly 100 constellations and a vast number of associated stars. It contains clear, historically oriented descriptions of stars, galaxies and clusters, all arranged in alphabetical order, and is profusely illustrated. The latest edition, which was updated by the author, Robert Burnham, to 1977, is published in three volumes by Dover Publications, New York. This is the best single work available on the stars and stellar phenomena and is likely to remain so for a very long time.

The Greek letters, *alpha*, *beta*, *gamma* . . . each refer to a particular star in the constellation. By convention, the accepted name of the star is given first (sometimes further identified by a Greek letter if in an open cluster), followed by the name of the constellation where it is to be found. Some examples will be given to indicate the diversity of names and certain other points.

Classification of star spectra

In the eighteenth century, and for most of the nineteenth, the planets were the focus of interest for astronomers in the West. To suggest that stars and galaxies were ignored would be an exaggeration, but they formed no more than a sort of backdrop to the planets. This may have been because of the absence of what climbers call a 'jug-hold', that is, a secure base for advancing beyond a tricky section. In other words, there were few organising principles, and there was no worthwhile general theory dealing with stars and galaxies – they were just there.

However, observation was laying a foundation for theoretical advance. The Frauenhofer–Kirchhoff–Bunsen study of the solar spectrum seemed a very promising research topic. It provided a distance scale for stars and galaxies, as well as measures of their velocities. In fact, most of what was known about the heavens was the direct or indirect result of spectroscopic analysis. (Light from the planets is not generated *in situ*. The planets are not self-luminous, so they reflect only light from the Sun.)

The breakthrough that led to modern cosmology was due to the individual efforts of female clerks working for about 25 cents an hour, and one female research associate who was paid a little more. They were

Williamina Fleming, Annie Cannon and Antonia Maury. Their classifications of stellar spectra was the springboard for a number of basic advances, including most of the subject-matter of modern cosmology.

The story begins in the 1860s when the Vatican made a belated *amende* to Galileo by setting up a Pontifical Roman College Observatory. This was an institute for astronomical research, staffed by Jesuits, and led by Father Secchi. He was the first to suggest that stars might be classified by their spectra, and he devised a fourfold scheme for the purpose, as follows:

Type 1   white or blue stars which show strong hydrogen lines in their spectra, for example, Sirius.

Type 2   yellow or orange stars with numerous spectral lines indicating various metals, for example, the Sun.

Type 3   orange to red stars, with wide bands and many fine lines.

Type 4   deep-red in colour with strong carbon absorption bands.

In 1864, after a five-year study of 4,000 stars, Secchi published the results of this analysis. This was the first attempt at a general classification of spectra, and was the starting-point of the Harvard Project, brainchild of the professor of astronomy, Edward C. Pickering.

Pickering was a somewhat eccentric academic, with a number of novel ideas. (He introduced students to physics at the Massachusetts Institute of Technology by having them do experiments in class in the laboratory – a first for the United States.) Angry with a lazy graduate assistant, Pickering told him that his Scottish housekeeper, Williamina Fleming, could do a better job, and then proceeded to prove his point by offering it to her at 25 cents an hour. At the time these were good wages for a cleaning woman but not for a research astronomer. However, Williamina was strong-minded and always demanded an explanation for what she was doing. Pickering found her argumentative and this did not suit him. So he hired another two dozen or so more amenable females to work on classifying the star spectra. Other, possibly envious, astronomers referred to the team as 'the Harvard harem'. These 'assistants' enjoyed some autonomy, and most of them did a lot of the thinking as well as routine work.

The first stage of the Harvard Project was completed in 1890, after five years spent classifying 10,351 stars of the Northern hemisphere in a revised Secchi-type, alphabetical, 13-group classification.

Meanwhile, in 1888, Pickering had appointed Antonia Maury as his full-time assistant to work on the second stage of his grand project. She was even more single-minded than Williamina Fleming, with – as Pickering put it – 'a passion for understanding' what she was asked to do. She abandoned Pickering's classification and developed an even more complex one in 22 groups, using roman numerals and insisting on having subgroups a, b and c. These served to indicate the appearance of spectral lines, and whether they were sharp or hazy, wide or narrow. In 1896 she left Harward and taught in private schools for the next 20 years. Belatedly her classification was acknowledged because it singled out supergiants from rapidly rotating stars. Her scheme appeals at first glance because it is complicated, seeking to come to terms with the real complexities of the problem.

The next assistant, Annie Cannon, worked closely with Pickering, helping to devise the definitive Harvard system for the analysis of stellar spectra. This was the legendary O, B, A, F, G, K, M scheme (immortalised in the student mnemonic as Oh! Be A Fine Girl, Kiss Me!). Working with glass photographic plates of the stellar spectra, Cannon devised a work plan that enabled her, with the help of an assistant, to classify three plates per minute. Between 1911 and 1915, she classified 225,300 stars. Annie Cannon was the niece of Henry Draper, whose widow had given a large sum in his memory to the Harvard Observatory. When the star catalogue was published in nine volumes in 1918, it was named in his honour.

There is no doubt that this work on spectra laid the solid foundation of fact for the theories of stellar origins and behaviours that are the glory of twentieth-century cosmology. These developments will be examined in the next chapter.

Recycling the ether – the Michelson–Morley experiment

There was a continuing anomaly in the accepted explanation of the velocity of light. In 1849 an ingenious experiment by Fizeau made it possible to measure this velocity even though it had the enormous value of 180,000 miles (288,000 kilometres) per second. (This is the maximum velocity of anything in the universe. In excess of this, matter would become involved with a 'singularity' – another way of saying 'we don't know what'. My own view is that it would be annihilated, being transformed into energy.) Fizeau's method was to measure the time it took a focused beam of light to be reflected over a great distance. A mirror in a designated spot, at a known distance from the light source, returned the signal to its point of origin. Interposed between the laboratory sending the ray and the one receiving it was a precisely machined, toothed wheel. This could revolve at various known set speeds. The speed was adjusted so that the outgoing ray of light was cut off by alternate teeth of the cog-wheel from the reflected ray. A micromeasure of the distance between the teeth was known, so a simple sum relating the distance travelled by the ray of light and the distance between the cogs yielded the velocity of light.

The Michelson–Morley experiment was more complicated than Fizeau's in its conception and design. After more than two thousand years of the 'ether hypothesis', Michelson defined the problem as follows: to measure the 'drag' of the ether on light travelling in the opposite direction to the Earth's rotation, relative to the 'push' given to light by the Earth rotating in the same direction. After years of discussion and preparation, the experiment was eventually carried out in 1887.

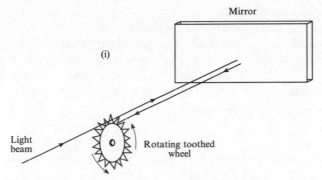

If wheel rotates too fast or too slowly the teeth will block the passage of the reflected beam. The speed of the cog at this point is the crucial measurement

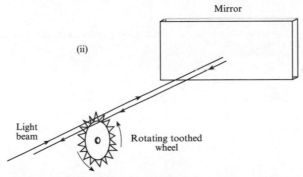

When the wheel rotates at the correct speed the reflected beam can pass between the teeth of the wheel. The speed of light can be measured by simple proportion

Figure 11   Fizeau's procedure for measuring the velocity of light

The rationale was simple and commonsensical. Imagine you are in a rowing-boat travelling upstream at 5 miles per hour. The river is running against you at 3 miles per hour. After one hour how much further upstream would you be? Obviously the answer is 2 miles. Turning the boat round to row downstream, what speed would you then be travelling at? Clearly, 8 miles per hour. But suppose you wanted to row across the stream, rowing at the same speed, how fast would you be travelling? The answer now is 5 miles per hour, but you would arrive 3 miles lower down the river for every 60 minutes it took you to cross over. Now instead of a boat on the river, imagine that we are observers travelling with the Earth at very great speed, passing through the 'luminiferous ether'. The question is, if that's the velocity of light moving in the same direction as Earth, what would be its velocity if it travelled at right angles or even directly against the rotation of Earth?

Michelson invented an instrument (called the interferometer) which could be used in tandem with a large telescope and spectrometer. Basically the new instrument consisted of two mirrors at right angles to each other. It could be used to transmit a ray of light for some known distance and back again. At the same time the ray was split in two by

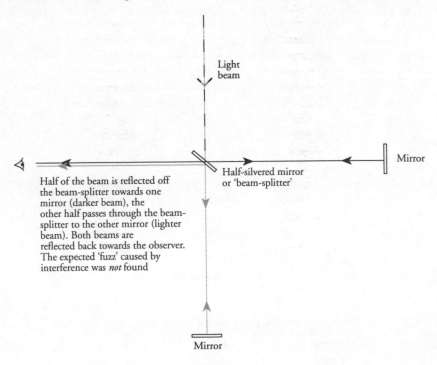

Figure 12 The Michelson–Morley experiment

using a mirror, of which one half was silvered and the other of plain glass. In this way, half the rays would pass through the plain glass and be diverted at right angles to the other half though both halves travelled the same distance. They were then both reflected back to the laboratory so that – at the point of origin – they could be timed. The vital question was, what difference would show in their time of arrival?

The short answer was – none. The two rays arrived simultaneously, just as though the ether did not exist (in fact it doesn't). The two rays, having travelled exactly the same distance, one with, and the other at right angles to, the 'ether stream', were expected to interfere with each other in ways characteristic of light rays whose wavelengths were not in phase. However there was no such effect. This meant that there was no difference in the measured velocities. The velocity of light was thus not affected at all by the 'luminiferous ether'. This was probably the most striking negative result ever achieved in the history of physics. As one authority put it, 'The whole notion of an ether as the carrier of electro-magnetic phenomena has been eliminated in contemporary physics.' This discovery was waiting to happen, because the properties of the ether were self-contradictory. It was said to be as rigid as steel, but to offer no resistance to planetary motion.

The end of the ether prepared the way for Einstein's revolutionary ideas of 1905. Before then there was just no answer to the question, what carries the light vibrations when waves travel from such places as the Sun to such places as Earth, or when you switch on a torch to go down a dark lane? The question was too simple, so the answer was wrong. Einstein's relativity theory placed the problem in a totally new context.

Michelson was the first American scientist to win the Nobel prize for physics, awarded in 1907. His final figure for the velocity of light in a vacuum (it varies slightly according to the medium through which it passes – air, water, oil, glass, and so on) was 299,774 kilometres per second. This was established in 1933. Michelson also obtained the most accurate measurement of the diameter of a star with his interferometer. In 1920 he recorded that alpha Orionis, better known as Betelgeuse, had a diameter 300 times that of the Sun. 'Beetlejuice', as it was commonly known, had its day of fame, the result being widely reported in the daily press.

Part II

Redefining the Universe

Chapter 5

Quantum Theory:
A New Look at Energy

The scientist . . . (does not have) the attitude of a man perched insecurely upon obscure and adventitious data. The world that is there has taken upon itself all the order, definition and necessity of earlier scientific advance.

G.H. Mead

A new beginning

At the beginning of the 20th century there was a ground swell of opinion that the accepted model of the workings of the physical universe (that is, the world outlook or paradigm of physics) was outmoded. The foundations laid by Galileo and Newton were under attack by leading thinkers such as Mach (1900), Planck (1900), and Poincaré (1904). The movement had started 50 years before, in what seemed the most certain branch of mathematics. Euclid's monopoly in geometry was smashed by three different kinds of non-Euclidean geometry, and Newton's notions of space and time became suspect. A whole new theory to displace Newton was sponsored by Riemann in the 1850s, and reasserted by Minkowski in 1904.

Physics, which after Galileo included cosmology, had long been a laboratory science which dealt with stationary and moving bodies, with force, and with the material side of events on Earth and in the heavens. Apart from the theory of relativity (1905–9), the two most momentous contributions to cosmology in the early 20th century were the Hertzsprung–Russell chart (1910), which gave the first rational classification of the stars (see chapter 8), and the theory that the universe was expanding unnoticed. This idea was first proposed by Slipher, and confirmed by Hubble. From about 1911 onwards, science stressed the enormous age of the universe as well as its continuing expansion. It was recognised that the galaxies, including the Milky Way galaxy, in which the Solar System is found, were retreating from each other as though they were mutually

73

repugnant. Slipher and Hubble were in control of the world's two largest telescopes, which gave them a critical advantage in thinking about the size, origin and evolution of the cosmos at this time.

Ernst Mach, forerunner of the new physics

Ernst Mach (1838–1916) was a vital figure in the new philosophy of science. His contribution was made in the late 19th century, but he set the scene for Einstein, a founder of the quantum and relativity theories. Mach was professor of physics in a succession of Austro-Hungarian universities – Graz (1864–80), the German University of Prague (1880–95) and Vienna (1895–1901). He became an Austrian senator in 1901. He had always been much more than an academic, and did valuable work on the human sense of balance, as well as in the philosophy of science. He was a trail-blazer in research, especially on the uses of experiment and of the thought process. His general philosophy now goes under the heading of operationism (or even Machism) – the view that all our hard-won knowledge in science is useless if it is not supported by laboratory operations. According to Mach, all our terms and theories must refer directly to, or be at a very short remove from, laboratory action, and must be promoted by numbers based on measurement. Such ideas were all part of the 19th-century revolt against metaphysics masquerading as science.

It should be stressed that Mach's interpretation of these key words – experiment, action, apparatus, knowledge and operation – was not a narrow one. It did not relate to a small back-room annexe where narrowly conceived work, on a par with licking stamps in an office, was done by grubby-handed artificers. Quite the opposite – it could span the furthest reaches of space and time. The work involved not only a laboratory full of apparatus, but also what were referred to as 'thought experiments'. In historical order, Mach influenced such people as Albert Einstein, the American physicist Percy Bridgman, who imported operationism to the United States, and the Vienna Circle, a school of professional philosophers described as logical positivists, which made a considerable stir in science in the 1950s. Among its achievements, it obliterated metaphysics as a quasi-intellectual discipline. It even stirred up action in England through A.J. Ayer, but got rather short shrift there, being considered not quite proper because it resembled science more than traditional philosophy. It was very Baconian in its approach to method. It did what the novelist James Joyce once declared was the main task in creative work, 'to cleanse the verbal situation'. This was prefigured in Mach's programme for physics, except that this was intended to abolish verbiage, and not – as Joyce did – proliferate new forms of it.

Mach was a world leader in the international community of scholars.

But our interest lies mostly in his strong influence on Einstein in the 1890s. Two Machian ideas, in particular, crucially reoriented Einstein's thinking. Apart from the most general principle, that 'nothing worthwhile can be said about any phenomenon unless the statement is capable of being tested in a laboratory experiment', Mach accepted the influence on scientific theories of our early environment, our instincts, our language and our thought processes. This was a great advance on the narrow empiricism of earlier 'positivists', and indeed of scientists in general. It was not intended to explain away disagreements, but to place them in context and thus help resolve them. But more important than his acceptance of the validity of 'instinctive' thought (always to be tested by experiment) was his recognition of the 'thought experiment'. He believed that the 'thought experiment', used correctly, was a worthy 'prologue' or 'anticipation' of empirical research. It could point the way to a hypothesis, or to a new way of testing it, or could serve as a context for understanding the results. However, he implied that to use a 'thought experiment' without the support of laboratory work pollutes physics with uncontrolled fantasy and subjectivism. Unfortunately, after Niels Bohr, cosmology became riddled with fantasies posing as thought experiments. However, used with the caution that observation and experiment are the key words without which theory and speculation must fail the acid test of objective reference, Mach's 'rehearsal method' is of great value.

As well as using data preprocessed by apparatus such as spectroscopes and telescopes, cosmology has in recent years begun to carry out actual 'experiments' in the cosmos, that is, on celestial bodies other than Earth. In 1992 the NASA spacecraft initiated an experimental programme in space on Mars, centred on the effects of outer space on animal life. Some quantum physicists have asserted that even thinking is an experimental interference with reality. But this notion is based on, or demands a belief in, psychokinesis, precognition and several other experiments in thought control of events. Leading experimenters in this field have sometimes been accused of falsifying their results, and partisanship has been the main feature of such reports since the origin of spiritualism in the United States and Britain in the 1840s. These speculations, and reaction to them, remind us of the ancient mapmakers' warning in certain blank areas: 'here be dragons'.

In connection with empirical research, Einstein can properly be described as a theoretical physicist who, by design, did very few physical experiments. He even avoided the regular laboratory training given students through class experiments. When he became famous, he was often asked about his 'laboratory', whereupon he would produce a pencil and say, 'This is my laboratory: I can work anywhere.' In short, his work consisted almost entirely of 'thought experiments'. The most famous of these thought simulations (they are now done on large computers) was carried out while he was still at school. He reports that after the physics teacher told the class that the velocity of light was 288,000 kilometres per second (or 186,000 miles per second), he spent the rest of the lesson – and indeed a lot of his later scientific life – trying to imagine what he would see if he were to travel to Earth down a sunbeam at this speed.

The philosophy expounded by Mach provided the context for Einstein's revolution in physics. In fact this revolution can be said to have begun with Mach's book on the history and logic of mechanics – the science of moving bodies. Mach maintained that only verifiable propositions were meaningful. In other words, what you see, or observe by means of your other senses, is what you present later, suitably processed, as a valid theory. The excess – anything which cannot be verified by experiment – must be taken to be *a priori* verbiage (or more bluntly, metaphysical garbage). It may be intensely interesting, and even amusing, but unless there is objective experimental evidence for it, it should be ignored. (We speak, of course, only about scientific matters. Ideology, opinion, fantasy, error and lies must remain as part of human political existence – but they aren't science. Creative use of ideology and fiction has a legitimate place in the arts, in debate and even perhaps in academic life, but not in scientific routines.)

'Economy' was a key word in Mach's philosophy. He believed that the principle of economy (called Occam's razor after the medieval theologian who affirmed it in Latin) was basic to science. Like 'ordinary' thoughts, scientific reasoning is a prime way of editing our report of events in the most economical way. In fact, even in science, there are no such things as 'extraordinary' thoughts. Science has no patent way of thinking which differs from the mental processes of non-scientific scholarship. It has a specific method, but this is true of any systematic inquiry, whether literary, medical, historical or other.

Science has two leading methods, or features, which single it out. It is directed towards understanding the external world by building on an existing and developing consensus, and it is self-correcting by means of experiment. In other words, scientific knowledge has an objective reference by which it tests its conclusions by experiment. Its language is quite precise, but it has no other special mysteries or spells with which to resolve problems of understanding. In science, magic, or even good feelings, are no substitute for hard work and hard-nosed experiments.

Bacon's *Advancement of Learning* (1605) anticipated much of Mach's general philosophy. Both Mach and Bacon contend that 'pure' experience, critically examined, will give us the source of our beliefs and understandings. For Mach, this meant our *total* experience, of libraries, education and work in laboratories. According to Mach, our experience, in this broad sense, must be recognised as the sole admissable (and self-sufficient) source of knowledge. For him, there is no special mystery, only mystification, that is the enemy.

Opening chorus – the trouble with science

Science is a territory which any citizen, male or female, can enter without benefit of passport, visa or payment of custom dues. The problem with traditional physics, which Mach labelled 'misplaced rigour', is an obsessional concern with appearances instead of with substance. This so-called 'rigour' is harmful, as it ordains that there are dues to be paid, a kind of admission fee for novices. For example, Plato is supposed to have had a sign erected at the entrance to his Academy, 'Let no one ignorant of geometry enter here!' This kind of exclusive-club spirit is foreign to the whole idea of science, let alone democratic principles. Accuracy and precise observations are essential to 'good' physics, but it is at least as important to generate a few creative ideas.

Einstein once said that it was better to start with a good theory (such as quantum theory or relativity) than to have our thoughts mechanically preempted by 'data'. For example, he said (privately) that Eddington clearly didn't understand relativity, or he wouldn't have been so nervous about the 'crucial' test he was proposing to make (1919): travelling halfway across the world on behalf of the Royal Society to test, during an eclipse, whether light really did 'bend' in the vicinity of massive bodies. Einstein's remark was light-hearted but it had a serious intent. (It has been suggested recently that the test was not really definitive in any case, since the error of measurement was about the same as the effect observed.) The whole episode lends support to Planck's and Einstein's position. The principles of relativity have since been repeatedly verified.

In fairness to Eddington, Einstein himself would have insisted that the observation *had* to be made, but he would have liked to have been spared the public-relations jamboree. Unlike religious creationists, who adopted a confrontation mode, the participants here were physicists engaged in an ideological revolution within their own subject. The person who initiated the new thinking was Max Planck. But these were only the beginnings, an opening chorus so to speak, of what was to come. And what followed was the most revolutionary period in the history of science.

Determinism and prediction

The classical view, established by Newton and a long line of earlier innovators, was most clearly expressed by in a statement by the Marquis de Laplace (1796), which has itself become classic:

An intelligence, knowing at any particular time all the forces acting in nature, as well as the momentary positions of all the things of which the universe is composed, would be able to explain the motions of the largest bodies and those of the lightest atoms in a single formula, provided his intellect was powerful enough to subject all the data to analysis. To him nothing would be uncertain; both past and future would be present to his eyes.

This statement was not, of course, intended as a progress report. It was pure hypothesis – and incidentally fails Mach's test. It is a programme, a statement of faith in the principle of causality, an announcement of final objectives, of aims and purposes. Metaphorically speaking, we move in a thought 'universe' where traditionalists, like everyone else, take their bearings from their everyday experiences. Many of us feel at home, even inspired by Laplace's declaration. It is clearly motivational, morale-building talk. We may dispute the fantasy of an omniscient observer, as a questionable residue of what Laplace called the 'God-hypothesis'. But the imaginary universe of which he speaks is familiar, and congenial, to scientists. It expresses an ideal, if unrealisable final purpose.

That said, it must be admitted that what Laplace proposes can now be achieved in a minor way in astronomy. For example, using my personal computer, I was able to observe a detailed simulation of the skies during the solar eclipse at the battle of the Medes and Lydians in 585 BC as predicted by Thales of Miletus. It takes about five minutes to feed the initial data into the astronomy program. The calculations use Kepler's laws to show the movements of the Sun, Earth and Moon as they would have been observed from ancient Athens. Ten minutes later, on the same computer, having watched the simulated Greek skies from the longitude of Athens, I was able to proceed to another eclipse as seen from Honolulu, even though this had not yet happened. (It took place in reality 10 days later, exactly as predicted on the computer screen.)

A scientist reading Laplace's declaration would be sensivitive to all the 'ifs' and 'buts' which soften Laplace's vivid metaphor. G.H. Mead's quotation at the beginning of the chapter (from his *Philosophy of the Act*) is apposite.

On the nature of light

The 19th-century historian Macaulay was addicted to the phrase 'as every schoolboy knows . . .' But not every schoolboy (or schoolgirl) knows that Isaac Newton's theories were pretty well accepted universally by scientists. True, a few physicists had mental reservations about some of his views, and claims were also made by various eminent individuals that Sir Isaac had misappropriated their ideas. (For example, there was a nasty dispute with Leibniz about who had invented the differential calculus. Newton hated being contradicted, especially on mathematical questions, and as a result of this dispute, he buried his discoveries under masses of paper, keeping them secret for many years.) However, there was a general consensus that his system worked, and gave an almost complete picture of physical reality, expressed as mathematical physics.

Christiaan Huygens, writing on light and gravity in 1669–70, was unhappy about Newton's views (as misunderstood at the time) on the nature of light. Certain findings pointed to the fact that light was made up of waves, not corpuscles as Newton claimed. These waves were thought to travel through the ether, eventually falling on some solid surface, or even coming in contact with small particles of matter, before becoming visible.

Taking the problem into the laboratory, it can be shown that, as in the case of water waves, the existence of one light wave does not preclude the presence of another. Two wave-fronts can be in the same place at the same time; they simply interfere with each other to produce characteristic fringe patterns which testify to waves in action. In other experiments however, light acts as though it were made up of particles or granules. Probably the most impressive example of this is where light falls on a metal surface. The metal reacts by throwing out electrons – tiny particles too small to be seen without special equipment. (The phenomenon is used in creating the television picture.) The electrons have clearly been 'knocked out' of atoms of the metal by photons (particles of light), travelling at enormous speeds. These act like high-velocity bullets, knocking the electrons out of the metal. This is known as the photoelectric effect, and the explanation given of it above was one of Einstein's first contributions to physics.

Some physicists, including Mach, refused to admit that there were atoms, because there was no direct physical evidence of their existence. Dalton's evidence (1808) was indirect, since it was a logical analysis of the laws of chemical composition (see pages 116–7). By contrast, Einstein's 1905 papers, one of which was on the photoelectric effect, and another on Brownian movement were accepted as the first direct evidence of the existence of atoms. Brownian movement (see Figure 13) concerns the zigzag motion of pollen grains in water; Einstein explained it as being caused by random strikes by atoms from all directions. (He was awarded the Nobel prize for physics in 1921 for this work, and not for his third paper of 1905, on relativity. The Nobel prize committee, like many councils, can be muddle-headed about what is really important.)

Physicists, avoiding the fine grain of Newton's argument, chose one or other side – waves or particles – in a rather contrived dispute on the nature of light. In fact Newton had proposed a kind of dual reality (a

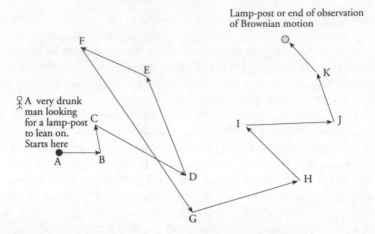

Figure 13   A 'random walk' and Brownian motion compared

'wave–particle') in his theory of 'fits'. He suggested 'a fit of this, then a fit of that', meaning that corpuscles of light travel through the ether in one fit, and then as waves in another fit. (In his later years, Einstein tried to estabish that all forms of energy behaved like this.)

It is important to recognise that a wave is a *disturbance of some kind which moves as a front.* By contrast, a force in Newton's system is transmitted to one body from another, in a straight line. The force acts at a distance without affecting intervening space (Newton's frame of reference is three-dimensional Euclidean space). There are several kinds of energy field which obey the same rules, being granular at certain times and wave-like at others. For example, cathode rays show this duality. They are stopped or diverted by massive objects, just like moving bodies or particles; they cast shadows like light, and don't go round objects as sound or water waves do.

Duality in wave–particles

Max Planck devised the quantum theory to explain his 'black body' radiation experiment. In this, a bar of iron (the black body) is heated until white hot and then placed in a black box to cool. (The colour black slows down the process of cooling, and the cooling process is observed through a minute pin-hole in the box.) Planck found that the heated bar successively emits each of the colours of the rainbow as the different ranges of temperature are passed through. He discovered that the rates at which these colours were emitted showed that the rays were given out, not in a continuous flow, but in distinct packets. These 'quanta' of energy (the Latin root from which 'quantity' is derived) differ in frequency as the rod cools, which explains the changes in colour.

This unexpected result – that light waves are transmitted in *packets* (as granules, corpuscles, quanta, call them what you will) and not in a regular *flow* – was more than disconcerting to physicists. It was as though you had been invited to hear a new piece of violin music written to show off the player's glissandos and vibrato – only to find there was no violin, only a piano. Violin glissandos can be approximated on a piano, but you miss all the frequencies and the notes which fall, as it were, in between the piano keys. Waves in nature are like the sounds of a violin or stringed instrument: the notes are not discrete but move continuously in sound and frequency. Grains, or quanta, however, are discrete, like the notes on a piano keyboard. They sound at clear and distinct intervals. As far as light is concerned, nature operates on the merging or continuous principle, as in a glissando or in the merging colours of the rainbow, but also on the 'exclusive' principle (illustrated by discrete pitches) as in photons.

Max Planck put his quantum theory forward in 1900 when Albert Einstein was only 21. But in 1905, while working as a Patent Officer in Switzerland, Einstein wrote four short papers (all published in the German *Annals of Physics*). Each paper made a major contribution, giving a new direction to physics. In one, he proposed to explain the photoelectric effect (see page 124). This account of the hitherto inexplicable and mysterious effect provided strong support for Max Planck's quantum theory. Einstein later introduced the word 'photon' to describe the granules (corpuscles or quanta) of energy which behave like material particles. The meaning of the word photon was later extended from light to other forms of particle–wave energy.

The Copenhagen manifesto

The quantum theory was taken further in an attempt to understand the detail of atomic structures. Niels Bohr was a Dane who worked in Germany with Einstein, in Cambridge at the Cavendish laboratory with J.J. Thomson, and in Manchester with Ernest Rutherford. One of the questions which may have bothered him was why atoms did not just finally collapse as they continued to lose energy by radiation. By all the laws of classical physics, the electrons ought simply to spiral in as this energy was exhausted, finally dropping into the nucleus to disappear without trace. (In fact it would be not quite without trace, for as the atom disappeared, rays at the wavelengths found in the spectrum associated with the element would be given out – whether calcium, radium, oxygen or whatever. It would serve as a kind of obituary notice.)

It is complicated to explain why this doesn't happen, and it is here that the inadequacy of classical physics is exposed. The point is that energy doesn't just leak out of atoms in a steady flow, like water lost from a leaky pail. According to Bohr's development of Planck's theory, the

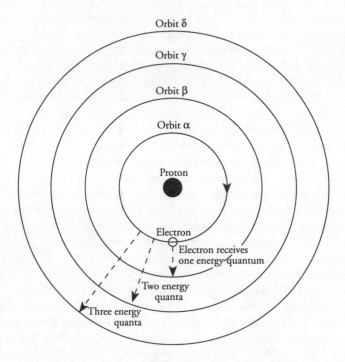

Figure 14   Bohr's concept of the atom

The hydrogen atom has one electron, at present in orbit α. This will jump to a different orbit if it has a donation of one energy quantum. More than one quantum will cause it to jump more than one orbit. Eventually it will become unstable in its new orbit and will begin to emit quanta of energy until it settles back on its original orbit. Its spectrum reveals these various structures.

process is more orderly, almost regimented. Energy travels and operates on the atomic level, in quanta, or packets. The quantum is not divisible into more minute portions.

Bohr's theory was that the life cycles of atoms are highly organised. Depending on the number of electrons circling the nucleus, and their distribution in rings, the atom is a stable configuration, not unlike the solar system. (The radioactive elements are exceptional exactly because they are unstable.) The configuration of electrons depends on the mass number, that is the total number of protons and neutrons (the latter having no charge) which form the nucleus of the atom. The atom is a three-dimensional system. The orbits of electrons pre-empt all available, 'empty' space. The electrons circle around the nucleus at set distances. These distances are established by the balance of forces of attraction between the positively charged protons and the negatively charged electrons, and the repulsive force between electrons. (Charges which are unlike, that is positive and negative, attract and discharge each other on touching, whereas charges of like sign, positive–positive, or negative–negative, repel each other.)

In Bohr's concept, while this process of electrons circling in their 'proper' orbits at colossal speed is going on, no energy is emitted or taken in. Then, for no reason at all – by some *unknown* and *chance* combination of factors – a particular electron is pushed forward, that is, 'selected' for promotion to a 'higher' orbit or demotion to a 'lower' orbit. 'Lower' and 'higher' here mean nearer or farther away from the nucleus. Because the nucleus is positive, whereas the electrons are negative, being nearer to the nucleus increases the danger of the electrons being too close to the protons and touching, which will result in the extinction of negatively charged electrons and a reduction of positive charge on the nucleus. This danger decreases the greater the distance between the positive protons and negative electrons.

To keep the balance, maybe one electron gives up a quantum of energy, which could pass to another electron which thus receives a quantum. Each electron will, as a result, 'jump' to a new orbit, one nearer and one farther from the nucleus. Thus there can be neither increase nor decrease of electrons in the atom. The element therefore won't change in its physical or chemical properties in the way that radium atoms do. (Rutherford had suggested the 'solar system' model of the atom: that it has a central nucleus which has a positive electrical change and contains most of the mass of the atom; the positive charge on the nucleus is balanced by electrons lying outside the nucleus.)

For Einstein, wiping out the principle of causality would mean the end of all science. He said as much to Bohr, commenting that having considered indeterminacy at one time, he had abandoned it because 'if all this is true [that is, random outcomes are a law of Nature in place of causal explanation], it means the end of physics'.

Bohr suggested to Rutherford that his new theory explained how the 'Rutherford atom' worked. More importantly, he demonstrated how the new theory explained the spectrum line of hydrogen, providing strong support for Rutherford's model. This brilliant *tour de force*, linking spectral displays to the innermost structure of atoms, is without a shadow of a doubt Bohr's greatest achievement.

We return to the phrase used above: 'For no reason at all'. This idea led to the parting of the ways between Einstein and Bohr in their thoughts about quanta. Einstein could never accept that physics was *forever and in principle* unable to explain how the fundamental process of 'chance' election or selection of electrons to 'jump' one way or another is decided. If the universe works on the basis of exceptionless laws, as Einstein believed, like all other scientists before Bohr, there is no room for fundamental processes which behave randomly. Politicians may throw dice when trying to decide the outcome of an unpopular or difficult decision, but Einstein talked as though the idea of God throwing dice were repulsive to him (as, no doubt, it is to anyone of even a moderate degree of sensibility). A more realistic model – one which leaves God and our feelings out of it – would be that of the universe as a massive pinball arcade, in which the celluloid balls in each atom are electrons 'jumping' from one pin to the next. But universal laws do not work like pinball machines. It is not a game of chance, where human luck, wishes, or intentions – as distinct from some human action, such as rocking the pinball table – can affect the outcome of atomic reactions.

Indeterminacy: Bohr vs. Einstein

Given four descriptors in the form of numbers – three for the location of a point in space and one for the time of day – we can pinpoint the position of an object at a given reference point at any given time. We can thus single out the nucleus of an atom with its specific structure of protons and associated neutrons (but as yet in Bohr's lifetime, only in a mental picture). However, what we cannot do, according to the quantum theory (and this, of course, directly contradicts Laplace), is to measure the velocity of a subatomic particle (minute body less than atom size) during a given interval and pinpoint its position in space during this same time interval. We can do it for the velocity and position of large celestial bodies, and with any small body moving on Earth, after any stated time. But we cannot do it at all for the velocity and position of subatomic particles at the same instant. These move with colossal speed by human standards. However, according to Bohr's theories, the failure has nothing to do with this problem. It is not due to lack of knowledge or equipment, nor to our poor understanding of the situation or the short-ness of the time interval. Rather, it is because the very act of measuring the velocity instantly alters the *position* of the atomic particle (when we measure its velocity at this point), or alters its velocity (when we measure its position at this point). We are trying to discover something that is unknowable, we are bringing to bear forces of unknown nature and extent which interfere with any expected result. It is though we were trying to predict the weather on 1 January of the year 5000 using our feeble abilities and knowledge of weather sequences.

The basic idea enshrined in the uncertainty principle is that when we begin to measure any one quantity of an electron (its location or mass or charge or momentum or energy or spin or velocity), 'the rest of the pack', so to speak (that is, all the other quantities) are immediately and randomly altered. It's as though when you choose a card, this is a signal for the rest of the pack, which your opponent has carefully arranged in order, to be shuffled. The question needs to be asked, why randomly? Do we have the records of such an experiment showing the various values for the 'rest of the pack' before and after? Presumably there must be half-a-dozen such changes in the course of any one experiment. In experiments on the theory of probability, for example, we can verify that specific events such as the scores made on specific throws of dice are not predictable. We know with certainty what all the *possible* scores are. If we are using more than one die, we can also say with certainty what the probability is for a particular score; we do this by listing all the possible combinations. But it is impossible to predict the actual numbers which will appear on particular dice.

According to one large group of scientists, even to *think* of measuring the position or velocity of the particle disturbs the atom and makes other measurements impossible. This is quite incredible. It comes from the same stable as the idea that you can receive a letter the day before it is posted. This attempt to confound our thinking about standard procedures and the accepted paradigm of physics is a central paradox of quantum theory. The paradox lies in understanding the motivation of professional scientists who assert that Kant and Hume were right all along. Arguing against causality, Kant and Hume asserted that the universe is ultimately unknowable even, some would say, by God. Not only is it unknowable: there were grave doubts expressed in Copenhagen as to whether the universe existed at all (except perhaps when someone such as God is there to notice it).

However, the claim that such views represented the latest findings of laboratory science at this time was premature. All the arguments reported so far referred to 'thought experiments', which had not actually been carried out. The existential problems were in the mind, not in the laboratory. At this time, the measurements which could not apparently be made had never been attempted. The argument was hypothetical, and the observations mentioned were part of some fantasy universe. (There are also recent examples of this. For instance, one text asks you to imagine that you are an exotic dancer in another galaxy and a million miles tall, so that you are visible from Earth. The question based on this fantasy was, did the audience in intergalactic space see you move your hand or your big toe first?)

The fairy-tale mode of writing popular science (of which the above is an example) is a stylistic innovation which creates problems for the reader. No clear distinction is made between analogical fantasies (sometimes drawn directly from, or derived from, Lewis Carroll) and real experiments. (Interestingly, in the 1920s and 1930s, *Alice in*

Wonderland seemed the favourite leisure reading of cosmologists, encouraged by Eddington especially when they attended international conferences.) There is no doubt that arguments became more serious and 'operationally based' after World War II, but before that there was a continuing air of frivolity.

Heisenberg: the new quantum theory

Heisenberg was the son of a professor of Greek. He was brilliant at mathematics but, having worked out the theory of turbulence for his doctorate (which he almost failed by reason of an administrator's stupidity), he came under the influence of Niels Bohr and decided that atomic physics was much more interesting. He was never terribly interested in philosophical questions, though he was attracted by Platonism, having studied the dialogue *Timaeus* in his high-school Greek class. Directly under the influence of Bohr, with whom he worked for a time in Copenhagen, Heisenberg accepted the idealistic philosophy of Bishop Berkeley (1685–1753) as expounded by Bohr. This philosophy was a down-to-earth version of Platonism (if any views derived from Plato can be called 'down-to-earth').

From 1925 to 1927 Heisenberg developed the 'uncertainty principle' (which has since been associated with his name though it had earlier been repeatedly stated by Bohr). This is the assertion we have already met: that it is impossible to measure both the velocity and the position of an electron at the same instant. It does not seem to have been noticed that the same principle applies if we try to predict which electron will be 'singled out' for expulsion from a radioactive atom in the elements radium or uranium, not to mention examples from the 'macro-world' of games of chance. It just can't be done unless you cheat. Nor was it clear what exactly was excluded by this principle. It was not the principle of determinism as such which was refuted, only our precise knowledge of a particular case, namely the position and speed of a particular subatomic particle travelling in orbit. As we have seen, Laplace's declaration leaves him open to criticism – this was not so much a thought experiment, even a bad one, as an uncontrolled fantasy.

To give an everyday example: we can toss 10 coins in the air, say 1,024 times, or have a machine do it for us. We can predict with considerable accuracy that about one quarter of the throws (256/1,024) will come up five heads and five tails. But we can never predict better than individual chance (in this case 50:50) which particular coins will fall heads-up or tails-up. Similarly, if we devise a truly random number generator, we can be very sure (but not certain) that each of the digits 0, 1, 2, 3, 4, 5, 6, 7, 8, 9 will come up an equal number of times in a very long sequence, of say 100,000 numbers. We can write a computer simulation program for the 10 pennies tossed in the air. We may illustrate the kind of pattern we would obtain by listing 200 randomly generated numbers. With the complete series of, say, 1,000,000 numbers each figure would appear 100,000 times (we can invite the computer to count them, it might take half an hour). This would illustrate the law of 'great numbers', which does not apply in full strength to our sample of only 200 numbers.

Rectangular matrix (20 × 10) of random numbers

1	6	8	4	7	0	4	5	1	6	7	5	9	2	9	9	5	6	7	3
4	0	7	9	5	7	6	2	4	1	3	5	9	2	4	9	3	1	3	7
6	8	0	4	6	7	9	3	3	4	6	9	0	7	4	1	8	3	5	4
5	7	4	1	3	8	9	4	1	8	5	1	9	6	9	8	3	2	0	9
3	2	3	5	9	5	9	5	2	6	2	1	8	7	3	1	2	1	9	0
9	8	3	9	1	5	4	2	9	5	1	6	1	5	4	9	8	2	5	6
7	9	4	8	3	9	4	3	9	9	6	2	7	9	6	7	9	1	1	7
2	5	9	4	1	6	1	7	2	6	5	3	4	3	9	8	1	5	9	5
3	8	6	7	2	3	5	6	0	7	2	1	3	6	7	1	6	0	2	3
5	7	4	1	3	8	9	4	1	8	5	5	1	9	5	0	6	8	3	0

Heisenberg, and nuclear physicists generally, were caught up in the simplistic debate about velocity and position. In fact it was not so much a debate as an extended session of academic name-calling. The quarrel was between the new wave of Platonists, led by Heisenberg and Bohr, who disregarded the effects of destroying the basic assumption of objectivity and determinism, and a group of older scholars, led by Einstein, who refused to acquiesce in this. Both sides failed to notice that Laplace, in addition to developing the science of celestial mechanics, had also advanced the unrelated, but now crucial science of probability. In fact, the two bodies of knowledge – celestial mechanics and the theory of probability – were his chief contributions to science.

Individual events, such as the 200 individual random numbers above, are inscrutable as far as the causes which decide which number appears where in the table are concerned. This problem is not insoluble, but it awaits someone clever enough to solve it (and it has been waiting a long time). But taken as a whole and as a system of random numbers (that is, a system that has no system), the table, in common with other probability distributions, is controlled by the laws of probability. (Once again, Aristotle must eat his words: he said that it was impossible to have such laws.) Our complete matrix of random numbers is determined by the rules of permutations and combinations, but we do not know how the individual numbers in a 'chance' collection will appear. The table above is an excerpt from the results of a computer program. Each single number is generated by a causal process, not by personal artifice or caprice.

We need dally no further over the vexed problem of objectivity. It is more important to note that Heisenberg advanced beyond these vague, metaphysical questions. Recognising that the quantum theory needed essential corrections, he borrowed an idea from Max Born (which he in turn had borrowed from Ernst Mach), that scientists should be concerned only with phenomena which were capable of being observed. The second basic change he made was due to Paul Dirac's revelation (1928) that we must abandon single, separate or individual numbers for the velocity, momentum and other measures of subnuclear particles because we cannot simply transfer the ideas of space and time developed from our macro-world into the micro-world of the atom. As far as theory was concerned, we had to do our thinking in terms of the new Einsteinian or Riemannian continuum of space–time.

Since it was now accepted by Heisenberg that it was impossible for the individual properties of electrons to be 'observed' seriatim, the best that could be done was to measure each of these properties (velocity, position, and direction of spin) separately, and in matrix form. In other words, we have to record each measurement separately, that is, one matrix for spin, one for velocities, and another for position. In so doing, we move back into the context where causality is an active principle which determines the distribution of many numbers, that is number patterns, and not just single numbers. It is not quite the old determinism, perhaps. But it is a causality behind seven veils, so the dance can go on.

This was not a matter for philosophical debate. At no time was Heisenberg as attached to the metaphysics of Berkeley as Bohr was. Although he promoted Plato's *Timaeus* as a cultural artefact at this time, Heisenberg, like Bohr, lacked the necessary mathematics (matrix algebra) and philosophy (epistemology) for the development of the new quantum theory, but he was a fast learner.

In the 1930s he was invited to Cambridge University by Peter Kapitza to conduct a seminar on his recent research. Kapitza was a visiting and distinguished Soviet scientist (he later headed the research program for Sputnik, which initiated space travel in 1957). He was an expert on powerful magnetic fields and on low-temperature research on gases. Paul Dirac, a doctoral student who attended Heisenberg's seminar, drew his attention to Cayley's work on matrices. Arthur Cayley (1831–95) had been professor of mathematics at Cambridge, where he developed matrix algebra.

The other crucial connection for Heisenberg was Albert Einstein. He was well known to everyone for his work on relativity, and equally so to physicists for sponsoring Mach's idea that non-observable items should be mentioned only in discussions designed to convert them into observables, or to declare them obsolete. (Einstein had done just this with absolute time and space, as Michelson–Morley had earlier done with the ether.) Putting all these views together Heisenberg came up with the new theory known as quantum mechanics. Dirac, who later became professor of mathematics at Cambridge, invented his own version of matrix mechanics.

Dirac's contribution

The changes described above did not come about solely as the result of academic discussion. The 'old' quantum theory of Bohr had reached an impasse. In his letters and conversations with friends at this time, Bohr anxiously described atomic physics as being in a state of crisis and pleaded for advice. His old version could not explain the spectrum lines of elements more complex than hydrogen, which had only one electron in the outer ring. Helium, with two electrons, resisted Bohr's attempts. The promise to read out atomic structures from spectrum lines worked only in a single case, and that the simplest.

The essence of the solution proposed by Heisenberg was to drop the idea of trying to draw a picture of the atom. Single numbers could not be placed side by side to describe the electron's velocity, position or anything else. For one thing, electrons were not viewed singly in the laboratory, but arrived in clouds. For another, quantum theory had no real interest in pictures even as visual aids, except maybe pictures of probability distributions. Instead of pictures, the scientist had to go after tables of numbers.

So, in place of single digits, the special numbers known as matrices were essential. These were square or rectangular groups of numerals (as in the matrix showing our table of random numbers on page 87). The essence of Dirac's solution was that the measurements had to represent what was happening not in one orbit, but in two. Each of the numerals in the matrix (two-way table) had to code a single event which had two aspects so to speak; thus each cell stood for two dimensions, one horizontal (for the atom giving an electron) and the other vertical (for the atom receiving an electron).

Now, a feature of matrix numbers when they represent a process is that sometimes they do not obey what are called the commutative laws of arithmetic. These are absolutely basic, at the foundations of arithmetic. We learn them not as abstract principles but as 'cases' or examples. Three times four (for example) is exactly the same as four times three; one-fifth of 10 ($\frac{1}{5} \times 10 = 2$) is the same as one-tenth of five ($\frac{1}{10} \times 5 = 2$). The commutative laws mean, quite simply, that, with simple numbers, in ordinary arithmetic, the order in which you add or multiply does not matter.

But the relations between the numerals in an array can be much more complex than those between single numbers in sequence, as our random numbers started out. These were arranged in a table (the defining feature of a matrix) mainly to save space, but they are actually a one-number-at-a-time string of numbers, one row of 200 events. We can arrange them in any way we like so long as our design registers the original order or sequence. A matrix, as in our discussion of quantum physics, usually represents a process. Some matrices do not obey the two most basic laws of commutation, those of multiplication or addition. This is why we refer to the matrix as numerals, not numbers. (The matrix is actually *one* number. By due process it can be collapsed into one single number which sums up all the numerals given. But this is a technical point, of no importance to the present argument.)

For a reason too complex to mention, matrices listing measurements on electrons sometimes show this non-commutative feature. They also give rise to the problem of indeterminacy. It is precisely in these non-commutative cases that we can't measure the velocities and positions of electrons at the same time. Nature proved to be more complicated than the older generation of quantum physicists had imagined. Even Einstein confessed that he was not an expert in modern mathematics. This particular problem was solved by Dirac.

Thus, the ancient metaphysical problems raised by Bohr and Heisenberg and discussed *ad nauseam* in the heyday of apologetics during the 1930s are quite general. Velocities and locations of subatomic particles are relevant only as special cases. Just as we don't agonise over which pennies in ten are going to be heads and which tails, so we don't concern ourselves with how many electrons can land on the point of a needle. The atomic model of the day was encumbered by excess baggage from past controversies. No one can be *certain* of every detail of the scientific solution, but then no scientist ever can be certain. This is the virtue of science – to eschew dogmatism while manifesting commitment. The basic fact is that the quantum theory, developed by Dirac and Heisenberg and combined with Einstein's relativity theory, forms the solid foundation of cosmological research. We still await the final synthesis, knowing that neither theory is an ultimate solution. We also know that Einstein's intuition was correct: the way to this knowledge is through a unified field theory. It is easy to say this, but it still requires the work of decades for it to be accomplished.

To sum up, here are three aphorisms from Dirac's lectures as a visiting professor in Australia:

the probabilities are all we need ... ;
it might be that Einstein will prove to be right [about causality] ... ;
quantum theory should not be considered the final form.

Some nonsensical interpretations

The academic world is by no means free from the confrontational idiocies which both support and bedevil other trades and professions such as politics, teaching, religion, medicine, the law, commerce, journalism and the manual trades. They may take different forms in each workplace, and the behaviours may differ, but basic attitudes remain the same. (Perhaps we should be grateful that our learned professions are still stocked by human beings rather than by faceless 'experts' who never admit to making mistakes.)

The quantum theory, because it overlaps with a number of areas of human concern, is one of these sensitive cases where people with a variety of opinions are in opposition to each other. As a result of its exponents playing to a variety of interests, numerous errors or idiocies have flourished, dating back to the 1920s. The fact is that the quantum theory is now closely wrapped up in the theory of relativity. In 1905 Einstein established that energy and mass, under certain conditions, not only interact but are interchangeable at a known rate of exchange. The real world 'works' in terms of this law. In fact, most of the phenomena affecting out lives – such as the Sun shining and making life possible in this part of the solar system, and the existence of the black holes which act as garbage collectors to the universe, disposing of the debris of worn-out stars and galaxies – are based on the fact that matter can change into energy, and vice versa. The discovery of this equivalence, and the fearsome demonstration of the change of matter into energy, may eventually be seen as the most important discovery made by human beings.

Einstein's laws are the foundation of universal activity. But they come to our attention only when bodies are moving at speeds which are a substantial fraction of the speed of light. Otherwise the travelling body is minimally affected. In ordinary circumstances (say travelling in a jet plane) the difference in weight of people and their luggage in motion, compared with when they are at rest, does certainly exist, but not so that you would notice. We have special problems, in short, but only with bodies moving at near-astronomical speeds, and with subatomic particles. When considering their mass, we need to use the new frame of reference provided by relativity, since mass is a function of their velocity. But there is no special or closed-off universe obeying rules created by some zany spirit. The universe was not created by Lewis Carroll. Neither is there a 'quantum universe' with special laws, whereby (say) a letter can arrive before you post it. There is one world only, with one set of laws, and these function, sometimes obscurely, but in accord with causality. We can show this unitary system at work and how the indeterminacy mode operates in terms of the size of the particle (see table below). We can measure the position and velocity of a large body without much difficulty. Uncertainty develops when we bring our measuring instruments to bear on microscopic or even smaller bodies. Similarly, it is true that the weight of an accelerating body changes, but at normal, terrestrial speeds the increase or decrease is quite imperceptible.

Table showing how uncertainty about position (p) and velocity (v) increases with decreasing size of bodies.

Nature of Body	Uncertainty in measuring p and v
(uncertainty increases)	(using instruments)
Cricket or tennis ball in flight	1 in 10^{34} (10 million billion* billion billion
A bacterium of scarlet fever	1 in a billion
Atoms in a crystal	1 in 100
Electron in orbit	more than 1 in 10

*1 billion = 1,000 million (10^9)

The table shows that we are more prone to error and uncertainty as we move from top to bottom.

Philosophy and physicists

Following the Greek model, traditional philosophy made absolute distinctions between factual knowledge and philosophical analysis. It argued that factual discoveries about nature had no bearing on or implications for our basic understandings of the universe as a whole; the existence of God; or the truth of our views about the nature of reality. This was the point Plato originally made about science, that the project of studying the natural world was misconceived, since such study told us nothing about 'the eternal problems' (as they were later described), meaning such things as the attributes of the divine or the nature of truth, virtue or justice.

Some backwoodsmen still maintain that to disagree with this is to commit the philosophical error they describe as 'type confusion' – a relatively polite way of saying that someone is so ignorant as to be invincible in argument. This is not as flattering as it might seem, for it is intended as a secret signal to sophisticates that the opposition is so stupid that they don't know when the logic of rebuttal has shown them to be wrong; that they are not fit to take part in arguments on the kinds of question referred to above, as it elevates them above their station.

Traditionalists who think this way were momentarily awakened from their dogmatic slumbers, and even became excited, about the Bohr phase in quantum physics. This happened when it became clear that Bohr was turning the findings of quantum physics about causality (or what purported to be the findings) against the Newtonian–Laplacian consensus. As we have seen, this model dominated physics for over three hundred years, and still does. It now became manifest even to antiscientists that 'facts', whether scientific or everyday, were never 'pure' and seldom 'simple'. Philosophical (and especially metaphysical) assertions are made within a context which includes a kind of sieve for straining off unwanted materials (for example, any scientific findings that contradict main-line metaphysical beliefs). This is true not only when science is in its reactionary phase, but even in its progressive phase. The label 'type confusion' is a convenient way to derail unwanted discussion.

Bohr resurrected the philosophy of Bishop Berkeley, subjective idealism, as the new epistemology of physics. He found support for this in Heisenberg's 'uncertainty principle'. In Britain, the early popularisers of quantum theory and relativity – Eddington and Jeans, both professors at Cambridge – sought to bridge the gap between science and religion. Ironically, they were attacked for their inadequacies in scientific logic and philosophy, as well as for their misunderstandings of Christianity, by Dean W.R. Inge of St Paul's Cathedral, London. Indeed, for many Christians, the subjective idealism of Bishop Berkeley, which was intended as an answer to scepticism, was even more unacceptable than determinism as the foundation of a real honest-to-God religion. Berkeley was an Anglican bishop of an Irish province, and hence a relatively free agent. (Church of England bishops can be tried for heresy, as Bishop Colenso of Natal learned to his cost in 1864, but this is rare.) Speaking generally, the 'type errors' condemned by traditionalists seem to be quite acceptable in argument used against science. Indeed, not to refrain from such arguments might be seem as failure of nerve in the 'battle' for religion. By contrast, scientists believe that an external universe exists whether humans are involved or not, and that questions of causality and the nature of matter and energy are clarified by empirical science. True, 'the God who plays dice' has become a new hero for some, but they should not be taken very seriously.

However, in the 1950s objectivity, together with the principle of local causality, were traded by many physicists for mindless 'spoon-bending' and 'psychokinetic' demonstrations. This was a symptom of the failure of nerve which seems to affect some scientists (especially physicists) during periods of great theoretical change. A number lose their way in pursuit of the crudest charlatanism. The science of physics was converted into a stage performance reminiscent of the craze for spiritualism whipped up by Sir William Crookes and Sir Oliver Lodge in an earlier generation.

Chapter 6

Relativity:
The Fast-Moving Universe

I have no particular talent, I am merely extremely inquisitive.
Albert Einstein

From Newton until the end of the 19th century, on the basis of his laws, physicists built a magnificent, precise and beautiful science which involved the celestial mechanics of the solar system, the theory of gases, the behaviour of fluids and of regular vibrations – in truth, a comprehensive system so robust and varied . . . [and] apparently so all-powerful that what was in store for it could hardly be imagined.
J. Robert Oppenheimer (abbreviated)

Einstein's relativity: phase 1 (1905–1911)

The person frequently described in his lifetime as 'the cleverest man in the world', Albert Einstein, was born in Ulm, Germany, in 1879. His father was a lapsed Jew, and Albert followed in his footsteps. He respected the Hebrew Bible and the history of Israel. Once, in reply to a telegram from an intrusive and boorish American Rabbi, he sent a prepaid reply to the effect that he believed in God, but that it was the God of Spinoza, unconcerned about human affairs, 'who reveals himself in the harmony of what exists'. He was a lifelong Zionist, but his views were radical and liberal. He rejected pressures on him to become President of Israel when this distinction was proffered. His comment on the Middle East is as appropriate today as when it was first made in *Einstein on Peace*: 'The ancient Jehovah is still abroad. Alas he slays the innocent along with the guilty whom he strikes so fearsomely blind that they feel no sense of guilt . . .'

Einstein attended elementary school in Germany and started on the secondary stage there. Until he was about nine or ten, he was considered by his mother to be dyslexic because of his difficulties in reading and speaking. Perhaps this 'dyslexia' was only the normal reaction of a sensitive child to the authoritarian, militaristic and possibly anti-Semitic climate of his school and social environment. He loathed the Prussian régime so much that, as soon as he could, he renounced his German citizenship. That was in 1896, when his family had already moved to Italy and he was at secondary school in Switzerland. He completed his second stage of education at the ETH (Eidgenössische Technische Hochschule) in Zurich. His curriculum included Greek, mathematics, German, science and one or two other subjects. He also took private lessons on the violin from age six, and music became an enduring hobby.

Einstein achieved world fame when the Royal Society of London sponsored a scientific mission to the Azores in 1919 to view an eclipse which verified his prediction that light would 'bend' under the influence of gravity as it came close to Earth. When the report of this mission appeared in the *New York Times* its headlines described Einstein as 'the man who caught light bending', an allusion to a popular music-hall song. (Newton had made a similar prediction much earlier, but the newspapers could not be expected to know that.) Overnight Einstein became more famous than a film star. This was when, in America, he was acclaimed as the cleverest person alive. At least one of his technical papers ('On the Unified Field Theory') was published much later, in full, in both the London *Times* and *New York Times*! This was done as an advertising gimmick. Needless to say, only a minute fraction of his readers understood it.

Einstein's fourth (1905) paper dealt with the energy equivalent of mass. Here he gave his famous statement of the relation, $E = mc^2$. In words, if we annihilate m grams of matter (a pound of tea weighs 454 grams), it will be replaced by light, heat, electricity and/or other forms of energy measured in ergs (a tiny unit in the centimetre–gram–second unit of energy) to an amount m multiplied by the square of the velocity of light (c^2). This is an enormous quantity of energy because of the extraordinary velocity of light. This energy appears instantaneously, as in the bomb dropped on Hiroshima in 1945. In the Sun a similar reaction goes on all the time and partly accounts for the heat and light which make life possible on Earth. According to estimates, it will continue for many more millions of years. Regulated and under control, as in atomic generating stations, this energy serves to generate electricity for domestic and manufacturing uses. Out of control, it produces a 'meltdown' (as at Chernobyl, in the Ukraine, and in various unpublicised cases in Britain and the United States). The annihilation of matter and its replacement by energy are the basis of many of the vast exchanges which underlie cosmic phenomena of all kinds. There is a converse reaction, where energy is converted to matter in particle accelerators, but this process is not at all spectacular, or at least is not observed on a regular basis as in the Sun.

Einstein's relativity theory arose from dissatisfaction with some eso-
teric failures of existing explanations for certain problems in light and
optics. It was anticipated in part by the German mathematician Riemann
(1826–66), and by the French mathematician Poincaré in a lecture given in
1904, and also was prefigured in James Clark Maxwell's equations for the
electromagnetic field which unified electricity and magnetism and predicted electromagnetic waves. The Michelson–Morley experiment (1881),
described in chapter 4, to detect the effect of the Earth's velocity on light
transmitted across the 'luminiferous ether', also cast doubt on the physics
current when Einstein was growing up. The results of this experiment
strongly indicated that there were no effects, and the concept was
dropped in science. The word 'ether', was nothing more than a catch-
phrase embodying another Greek myth, to be buried along with the
unicorn and the phoenix.

The Michelson–Morley experiment created a minor crisis in physics.
But, contrary to report, it did not influence Einstein towards relativity; he
later said that he didn't even know about it. However, it did make the
theory more acceptable to others. A more difficult problem concerned the
planet Mercury. Every year, it arrived exactly at the point it should do,
but in advance of the time predicted. The error was cumulative, and could
not be explained. (The place in question is called the *perihelion*, meaning
the shortest distance of the planet from the Sun during its trip around the
Sun.) The annual failure to conform was called the 'precession of the
perihelion of Mercury'. It was rather disturbing because inexplicable.
Einstein's theory of relativity provided the explanation.

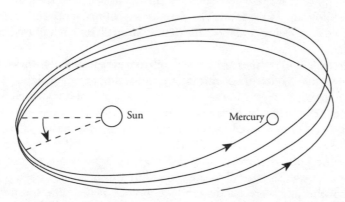

Three and a bit orbits are shown of Mercury's circuits around the Sun. The
angle shows the discrepancies in the position of the nearest point of
Mercury's orbit (the perihelion) to the Sun and how this discrepancy
increases over time

Figure 15   Precession of the perihelion of Mercury

Whether light was made up of waves or particles was another problem. Newton left some doubt about this question, though his views were clearly expressed. There was yet another unresolved question: was it true that matter could be destroyed by due process and transformed into an equivalent amount of energy? Radioactivity was a feature of certain metals, like radium, which break down naturally. This seemed anomalous in Newton's tight little universe.

A few physicists like Mach also drew the line at Newton's absolute time, an idea taken over from Aristotle. Time differs from place to place because every natural society has adopted the Sun's changing position in the sky as an indicator of time passing. Newton, however, believed that each time and place were preordained. In other words, there was an absolute value of time which was established everywhere; our clocks and maps, if accurate, recorded eternal values set up at the Creation. According to this belief, there was a location somewhere, where latitude and longitude began, and at that location was a clock which kept the exact time. It was as though Greenwich Borough in London had existed (in the mind of God, perhaps) with the absolute latitude and longitude of zero, as well as an error-free clock, recording absolute or God's time. So it had been for all eternity.

Other places and other clocks, suitably adjusted for time, latitudes and longitudes according to their geography, were also thought to be eternal. As people travelled across the world and emigration became more common, times and locations were subject to human whims and convenience (with the widespread use of daylight-saving times), so that today we recognise that these constants are similar in origin to such units as the metre, kilogram and second. They are merely human conventions; no one place, no one being and no unit being 'special' or having any 'natural' priority.

Einstein's 1905 paper on special relativity repeated the message first delivered by Galileo. It is one of the most important insights we have about moving bodies. Moving at a constant speed, and observed within a frame of reference which moves at the same velocity, bodies appear to be at rest. Galileo illustrated the principle in terms of a passenger on a ship. We all know, but don't normally think about it, that if there is no rolling motion or you are down below decks, there is no way you can tell the ship is moving. Things fall to the ground in a straight line, not a parabola, just as if the ship were standing still. You walk at the same speed whether going fore or aft or diagonally. You don't have to speed up, slow down or change direction to compensate for the forward or backward motion of the ship.

The same is true for passengers on a train moving slowly at uniform speed, as it cautiously leaves or enters a station. The platform and any stationary train seem to be moving, but the train we are on seems to be at rest. This kind of situation was what Einstein had in mind when referring to the special theory of relativity. (But this was the easy case, already described by Galileo.)

The general theory of relativity deals with an observer in an accelerating frame of reference, such as a falling elevator or a person jumping out of a burning building, both being examples of free fall under gravity. This theory had to wait another eight years before it was discussed by Einstein, for such cases are much harder to fathom. In a 'feet-on-the-ground' frame of reference, the Earth moves at a uniform velocity. But in a runaway lift, say, or in a bungee jump (when you are crazy enough to pay to jump off a cliff with a strong rubber tie around your ankle), matters are quite different. The jumper in free fall is constantly accelerating (the acceleration due to gravity, $g = 32$ft, or 9.8 metres, every second).

During the early phase of the relativity theory, lasting until 1911, Einstein persuaded physicists and cosmologists that there was no such thing as absolute time or space, and that the only absolute constant in the whole universe was the velocity of light. This always has the same value, and in a vacuum it is, in round figures, 186,000 miles (approximately 300,000 km) per second. The individual rates at which each of the radioactive metals, such as radium, uranium, polonium or plutonium, disintegrates also never varies. It is the same whatever the temperature or pressure. Even united with more ordinary materials (for example in radium chloride, a salt where the radioactive element is joined to an acid radical), the rate at which the radium disintegrates stays the same. Likewise, the velocity of light is quite unaffected by surrounding changes except when travelling through glass, oil, water, or some medium other than air.

Relativity phase 2: the established principles

(1) Relativity has to do with moving bodies when observers are themselves moving. If bodies are at rest, or the velocity of the observer is small compared with the speed of light, corrections made to the observed velocities are themselves quite small but increase from zero velocity. At half the speed of light and above, the Einsteinian corrections are substantial.

But the universe is not, of course, a Russian doll which conceals other dolls beneath its exterior, or in this case other worlds. There are certain key points, for example changes of state such as water boiling, or the velocity of light being 'capped' at a limiting value. But just as there is no quantum world, so there is no relativity universe either. There are no special microscopic (atomic) universes, nor any special astronomical universe. There is just the one universe we all live in, with habitats for microbes and space for galaxies and for bodies travelling at unimaginable velocity. But the microbes and the galaxies interact with us according to a single set of laws, albeit from a distance, even an astronomical distance.

Much confusion has been introduced by authors such as J.B. Priestley, writing on time and anxious for a new twist, by science fiction writers and by preachers looking for a new gimmick. Popularisers eager to exploit sensational analogies have also done much to obscure the stark truths of relativity. The tendency is an old one, established long before Bernhard Riemann and Albert Einstein attempted to extend our horizons by talking about new dimensions of space and time.

(2) Clerk Maxwell (1860) believed that light was a kind of energy disturbance, electromagnetic in nature and origin. Einstein believed that gravity was similar to light in being a field phenomenon, with gravitational waves in place of light rays; the universe was full of invisible gravitons, which were constantly being replenished, just as light is full of photons. This summary does not do the idea justice, but even so, the same questions arise as with the theory of light. Are gravitons waves or particles, or both? Does the gravity field behave like the magnetic field? Are there 'neutral points' where opposing 'forces' (gravity clashing head on, but working in opposite directions) balance so that no movement takes place 'under gravity'? Is it possible to map the local gravity field in this room, for example, as we can the magnetic field? In brief, how, so to speak, do we get a 'handle' on gravity as a field of action?

There are many questions but few answers. We can make sense of water waves and even light, but what about gravity waves or gravity-wave particles, which can't even be seen? Furthermore, having to translate all our ideas from the 'force-acting-at-a-distance' Newtonian ways of thought into a new 'space–time-continuum' way of thinking is like being lost in a foreign country where we don't know the language. Einstein himself tried for twenty or so years to solve the key question of the unified field theory. Since 1955, scientists have been able to show that gravity waves exist, invisible and very small.

(3) Clerk Maxwell revived the Greek theory of the 'ether'. As he understood it, just as water carries water waves and air carries sound waves, so the ether medium carries various kinds of invisible energy waves – for example, the 'fields' (or 'forces') having to do with light, heat, magnetism and (later) radio. The Michelson–Morley experiment cast doubt on the existence of the ether. But abandoning the concept at that time would have left most people with a mental blank. We need the word, and even the idea of the ether (albeit temporarily, until we can all speak the language of fields), just as the ancient Mayans of Central America needed the image of a procession of day-gods who transported

time. The Mayans believed that this caravan of deities walked in line to deposit their 'day-packets' at a day-station on the way, before moving on to pick up their next load. This procession maintained the universe, our existence and time, and so ensured the survival of the human race. Our common language needs similar images, such as 'the ether', to interpret abstract ideas.

(4) Many experts, including the Dutch physicist Lorenz and his Irish counterpart Fitzgerald, tried to save the ether concept from oblivion. They suggested, separately and on the basis of a 'thought experiment', that wooden measuring rods travelling at very high speeds were also subject to foreshortening in the direction of motion (if instead a rod travelled sideways it would end up as a straight line of wood). For example, a metre stick starting from rest and increasing its speed would contract until, at nine-tenths the speed of light, it would be only half a metre long. Its mass would increase by 2.3 times, so the wood would be more than four times as dense. At the speed of light its length would shrink to zero. (Above that speed, which defies the imagination, we and the metre stick might just be involved in a 'singularity' from which we might not recover. Would the metre stick reappear as it slowed below the speed of light, like Lewis Carroll's Cheshire cat in reverse? I wonder.)

Similarly, a clock would go slow, eventually stop and proceed to disappear as the velocity increased. This is known as time dilation. The Lorenz–Fitzgerald contraction means that the object being measured along with the metre stick which is measuring it would contract, each by the same amount. So the Michelson–Morley experiment 'preserved the appearances' in Plato's phrase; like Alice, we would not notice any difference in a universe where everything (including ourselves) travelled at immense velocities but below the speed of light.

This bright idea was discussed *ad nauseam*. Unfortunately, like all the ideas put forward to rescue the ether, there was no way of testing it experimentally. More recently, there have been experiments with clocks and other materials in elevators in very tall buildings. These seem to support some of these scenarios. There is also some evidence from high-velocity subatomic particles. But, by Mach's criterion (see page 46), this area remains highly speculative.

A comparable case is the so-called 'twin paradox'. The game here is to pretend that a clock is travelling at very high speed, almost at the speed of light. You also need a pair of identical twins and two error-free atomic clocks which record days, weeks and years and give the time of day. One twin stays at home for, say, ten years, living a normal life, always travelling in such a way that he observes the speed limit. He takes care his clock is not damaged or stolen. The other twin is sent off to another galaxy at very high speed, with his clock. He is brought back at the same speed for a family reunion. The clock which stayed at home with his brother is found to be very fast compared with the clock which travelled to some outer galaxy in space. We assume that this discrepancy in the passing of time has also been a feature of the twins' metabolism over this period. So one twin will be physically older than the other, and will look it.

This is an expensive, if ingenious way, to test Oscar Wilde's story *The Picture of Dorian Gray*. But the question is, would it work? Unfortunately, the answer is no. A precondition of the experiment is that there be no change in the inertial conditions from start to finish. In this case, our curiosity which leads us to bring the extra-galactic twin home violates this condition. To come back to Earth, the twin has to make a 180-degree turn, or to plan the trip so that it takes in a wide circle. In either case, inertia alters markedly as movement direction changes. What we have, in fact, is another thought experiment; and there is no way to test it in reality which does not contravene at least one of the necessary conditions outlined above. There is no true paradox here, merely an idea for a Hollywood film-script. Einstein's adolescent thought experiment is a different matter: it led to the fundamental equation linking mass and energy. Einstein's thought could be verified by action, the testing of related variables and processes. It was real, not entirely fantasy.

(5)   Einstein's recognition of the equivalence of mass and energy was perhaps the greatest contribution that relativity made to human knowledge. Earlier, we met the equation that declares that matter can be converted into an equivalent, and very large, amount of energy. But even the different kinds of subatomic particles can be converted to energy, usually gamma rays (see chapter 7). Conversely, particles are also created in the course of nuclear fission (splitting), and by fusing nuclei in high-speed particle accelerators. Einstein gave us the formula showing how the mass (that is, the amount of matter in a body) is increased or decreased as the velocity increases or decreases. This particle moving at very high speed has more mass than when it is at rest. In exchange, a large amount of free energy is liberated into the universe when mass disappears in certain reactions.

(6)   Hermann Minkowski, who taught mathematical physics to the young Einstein in Zurich, pointed out in 1908 that the interval between two events shared some of the characteristics of distance. He was starting from Euclidean geometry, as taught to schoolchildren, but was moving away from it. Euclid related events in the real world to three axes at right angles to each other, such as the corner of a room at floor level. Any point

in the room can be located by measuring its distance from these three axes (the three lines where the floor and the two walls join).

We could set up a camera with cross-wires focused on the corner and keep a diary of what we observe in the small space visible in the course of the day. Suppose that we set up a security system focusing on a safe which is kept under 24-hour surveillance. We can write up our report in terms of *four* dimensions, not three. They are distance up from floor, distances along one wall and along a second wall, and the time. If we draw lines as distance markers on the walls and floor and provide ourselves with a clock, this will enable us to make our observations more easily. The point is, that time (one dimension) and space (three dimensions) *taken together* are our usual way of specifying events. Einstein adopted Minkowski's time–space continuum in developing his relativity theory. Here space and time merge in one integral manifold: a continuous multifaceted integral dimensionality – a four-in-one, so to speak.

Relativity phase 3: gravity and the time–space continuum

Einstein, in his paper on the special theory of relativity, dealt with the electromagnetic field. His paper on the general theory of relativity considered the application of relativity principles to the gravitational field. The essential difference was that, in the latter, bodies were accelerating and not moving with uniform motion. Since Newton, the traditional view was that forces such as gravity, magnetism or static electricity act at a distance between two bodies (either magnets or charged bodies), across empty space, as we might say. They do so in accordance with the law of inverse squares, such that the greater the distance, the more the force drops off, not with increasing distance but, very rapidly, with its square.

The very concept of a force acting at a distance was queried at the time by Huygens and others, and later, in the nineteenth century, by Oersted, Faraday and Clerk Maxwell, the discovers of electromagnetism. They explained electric and magnetic fields as states of tension of a specific kind which spread in all directions, almost instantaneously (as we now know, at 186,000 miles (300,000 km) per second), whenever and wherever you might switch them on (we do this whenever we switch on any household electrical appliance). The fields act much like gravity: the electromagnetic law and Coulomb's law for charged particles are also laws of inverse squares. These similarities were enough to make Einstein conjecture that some simpler law should cover both. He devoted more than thirty years to this task and produced some ten different answers, all of them wrong. His first paper on the subject was in 1923; he died in 1955 with the problem still unsolved.

Einstein once remarked that the laws of science should be so simple that they can be expressed in a way that a child can understand. He himself was often asked to prove his point, and to explain the nature of his discovery to newspaper reporters. His ability to explain things clearly is illustrated by a story he is said to have told originally to his 7-year-old son. A blind beetle (he said) had somehow fallen onto the world globe Einstein kept in his study. When Einstein first noticed the beetle, it was at the equator, and as he watched, it crawled due north in a straight line as far as the North Pole. At least, the *beetle* thought it was crawling in a straight line, but Einstein could see that, since it was on a globe, the route was not really straight but curved. When we ponder the latent meanings, this is about the clearest explanation of the geometry of space–time that one could imagine.

The geometry of space–time

Bernhard Riemann (1826–66) was a creative and versatile German mathematician who produced a host of ideas about the geometry of space. He described a geometry quite strange to the traditional view and in direct contradiction to prevailing ideology. Mathematics at the time tried to teach us that Euclid's representation of the world was based on the realities of everyday life. Riemann said, in effect, that this was a geometry for blind beetles, and that we should stand up and see for ourselves. He presented us with a 'universal' (that is, general and abstract) geometry, which became the basis of new thinking about space.

In Riemannian geometry no lines are parallel to other lines and all straight lines are all the same length. The angles of a triangle always add up to more than two right angles. Similar polygons (that is, figures with the same shape but different sizes) are impossible – they don't exist. Space can be of two, three or more dimensions. Space and time are not separate from each other, but are integrally connected in a single manifold. Since they are continuous, we need to learn to speak of the 'continuum' of space–time and of the fourth, fifth and even higher dimensions.

Riemann's universe was not topsy-turvy, but topological. To present it 'neat', as has been done here, may make the ideas seem ridiculous. But this is because most of us don't see the universe as it is, but from the Euclidean, 'blind-beetle' standpoint. Riemann's geometry takes account of the fact that we live on a vast sphere, but it prepares us to deal with any kind of space which might be encountered or even imagined.

Contrary to our expectations based on the ancient geometry, in Riemann's geometry, as in real life, 'things' can often be stretched like rubber. Putty rings can be twisted until the hole almost disappears, and solid coffee-cup shapes can be squashed flat as though wet clay. Any material can change its shape and configuration. This kind of geometry, known as topology, was beyond Greek intellectuals (such as Euclid) because they worked not with things but only with ideas. In his book *The Foundations of Geometry* (1850), Riemann formulated rules starting from the simple relations between points and then taking off for a continuous, filled space – one, as he puts it 'without distinction between gravity, electricity, magnetism or heat'. In his geometry, unlike Euclid's, a straight line was not necessarily 'the shortest distance between two points'. Together with Einstein's demonstration that light does not always travel in straight lines, it finally and conclusively pulled the rug from under Euclid's version of the universe.

Gravity waves

In Einstein's theory, the nature of gravity and its mode of action are closely bound up with those of light. In fact, the more one considers the matter, the clearer it is how far Newton's synthesis of physics missed the mark. This is not to denigrate Newton; it merely points to the condition of learning in his day. What is needed is an *integral* solution of the problems of the universe, that is, a general law, or laws, to cover all forms of energy, all types of motion, and give an explanation (in terms of physics not philosophy) of the nature of matter, of all energy fields such as light, gravity, magnetism, electricity, heat and so on. In search of a solution, Einstein gradually withdrew from the scientific fray, devoting himself to try at least to unify the theory of light and gravity. But after 1945, he increasingly tended to be regarded as 'yesterday's man'.

The fact that light was a form of energy, and had a dual nature as corpuscles and waves, was the key. Referring to the waves, we speak of 'rays of light', an everyday recognition of this field characteristic. Of course, when the Sun is obscured by clouds, it is still daylight, even though we can no longer see the rays except when they pass through breaks in the clouds. When this happens, we can *see* that light is made up of rays, and may be a wave motion. But gravity waves or rays cannot be seen in this way: although we know that 'the field' exists as gravity acts, and can be seen to act, wherever we may be.

Einstein began to study the gravity field as early as 1916, before the brouhaha developed over the indeterminacy issue (see page 84). In one experiment he used a dumb-bell, rotating it so that the rounded ends cut rapidly across the gravitational field. This experiment was modelled on the induction experiments which Faraday had carried out at the Royal Institution in 1832, to generate electricity. The Earth is surrounded by a magnetic field which runs roughly from north to south. Faraday's experiment was to rotate a coil of wire very rapidy around a bar magnet so that it repeatedly cut across the 'lines of (magnetic) force'. This 'induced' a current of electricity to flow in the wire. It could be drawn off and 'stored' in a battery for further study or for practical use. This discovery, which we now take for granted, features in about 90 per cent of our everyday lives.

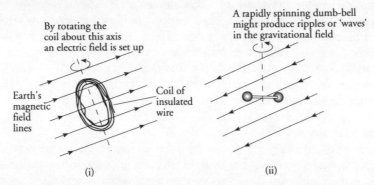

Figure 16   Faraday's induction principle (i) and Einstein's expectation of gravitational waves (ii)

By rotating the dumb-bell across the gravitational 'lines of force', Einstein hoped to single out gravity forces in the field, and reveal how they work. He hoped to give rise to some phenomenon, similar to light or magnetism, that would, so to speak, give a handle on gravity as an energy force.

Gravity should not to be thought of only in terms of its local, purely parochial effects, such as causing apples to fall from trees. There must be other local effects just as revealing (and useful) as the electric current in Faraday's experiment. It was expected that gravity waves, like Faraday's electromagnetic ones, would travel at the speed of light. Electric appliances such as light-bulbs, vacuum cleaners and the like, all serve to drain away the electric energy generated by induction in electricity stations. Perhaps, thought Einstein, gravity too might carry away dynamic energy from other kinds of system, such as massive bodies (like planets revolving in their orbits), or small bodies (like apples falling from trees). The basis of these expectations is the second law of thermodynamics.

We can think of the forces of the universe as being strong or weak. Gravity is classed as a weak force, so we can anticipate the need for 'multipliers', or special circuits, as in electricity, to accumulate and distribute this force in local generating stations. Looking further ahead, we can foresee the production of anti-gravity devices and the application of 'weightlessness' devices for the disabled or for lifting heavy objects. Earlier generations of naturalists puzzled over the question of how a female elephant could support the enormous weight of the bull elephant during mating. One brilliant sugestion was that copulation took place only when both were partly submerged in a water-hole, since this would reduce the weight considerably – a practical application of Archimedes' principle. Someone else suggested that, to find out the answer, they should go and see it happen. (This parallels Einstein's insistence that it was not enough just to talk; physics also involved looking, doing and testing.)

Slipher, Hubble and the expanding universe

New thinking about the real size of the universe and the almost infinite number of stars and galaxies dates from about 1924. At this time, the American astronomers Slipher, at Flagstaff Observatory, Arizona, and Hubble, at Mount Wilson Observatory, California, discovered that some nebulas previously thought to be inside the Milky Way, and so relatively close to Earth, were in fact quite some way out, in some far-off and independent galaxies. This revolutionary thought not only implied a much bigger universe, but also led to the discovery that these external galaxies were travelling away from Earth with colossal velocities. Slipher obtained the values from the Doppler effect at the 'red-shift' end of the spectrum. His findings were the first independent observational evidence that the universe was 'expanding'.

If anything, Hubble's contributions to our understanding of the universe overshadow even Slipher's. He was in charge of one of the two biggest telescopes in the world (the other was Slipher's), and worked closely with Einstein in correcting his calculations and putting right what Einstein described as 'the biggest mistake' of his life, namely the inclusion of the so-called astronomical constant in one of his equations.

Hubble developed the first proper classification of galaxies (see chapter 4). In 1929 he discovered what became known as Hubble's Law: that the galaxies are retreating from us, and from each other, at velocities directly proportional to their distances apart. The farther they are from each other the faster they are in retreat. This is an extraordinary finding. The distances were calculated by the standard method described earlier as the 'beacon' system, and the velocities were calculated by the 'redshift' spectrum method. Hubble's constant is calculated by dividing the distance of the external galaxy by its velocity of retreat. It is the fact that it is the same value in each case which confirms that the universe is expanding rapidly. The velocity of expansion is in line with the distance apart of the galaxies. It is as if the universe were a balloon with spots on its surface. As we blow up the balloon, the spots move farther and farther away from each other.

In 1929 Hubble established that the redshift as measured indicated a velocity of separation of about 980 feet (300 metres) per second. But later, when he found the galaxies to be more distant than previously calculated, he revised this estimate to about 45 miles (73 km) per second. The reciprocal of Hubble's constant (10 thousand million years if we grant him certain assumptions) is the best estimate of the age of the universe. This is not in conflict with our figure for the age of the Earth; on the contrary, it supports the vast age of the primal PreCambrian rocks.

Einstein accepted Hubble's view that the universe was expanding, and modified some of his relativity equations accordingly. He worked at Hubble's California observatory for about three years, checking his observations and discussing their interpretation. He was then lured away to a post at Princeton Institute of Advanced Studies by Abraham Flexner, who specialised in 'stealing' the best brains for his establishment.

Gravity as a field

Under the new dispensation of relativity, gravity was seen not just as a force acting between bodies anywhere at a distance from each other in obedience to Newton's law, but more broadly as a *field* very like the Earth's magnetic field. As we know, this stretches out in all directions to infinity, being a state of strain or tension which manifests itself in ever-decreasing magnitude as we move away from the magnetic poles and into outer space. Of course, when it encounters other magnetic fields, even the tiny one created by a pocket compass, it is distorted or twisted locally, as sea waves are distorted when they come upon a rocky coast or coral reef. In other words magnetism, like other forms of energy, manifests itself in two forms: as a force of attraction between unlike poles (north-seeking and south-seeking); and as a force of repulsion between like poles (both north-seeking, or both south-seeking). Static electricity, such as builds up in storm clouds and is discharged as lightning, has similar features. But static, unlike magnetism, is evanescent. Current electricity and radiant energy in its several forms, like hydraulic energy, exist both as waves travelling through the appropriate medium, and as particles – solid (or, in the case of water waves, liquid) matter. Does gravity show similar features? Einstein thought that it might.

As early as 1600, William Gilbert, in England, recognised that the Earth acted like a huge magnet, affecting compass needles everywhere. Faraday showed in the 1830s how magnets could be made by passing a current through a coil of insulated wire wound round a bar of iron (called a solenoid). These magnets' fields can be placed in opposition to each other, or in cooperation with each other or with the Earth's field. The magnetic fields and their interactions can be seen by sprinkling some iron filings over a piece of paper placed on top of magnets laid in various directions. The filings arrange themselves along the 'lines of force' in the horizontal plane.

It was only in 1960 that Einstein's general theory of relativity was taken out of storage and recycled. The Grand Unified Theory of the

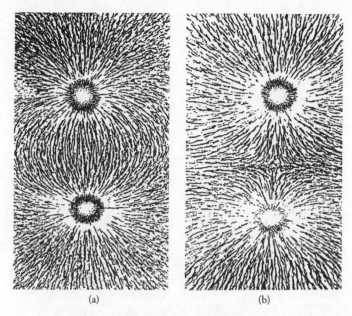

(a)　　　　　　　　　(b)

Figure 17 The magnetic field lines produced by two unlike poles (a) and by two like poles (b). In (b) there can be seen a neutral point at which the field lines diverge. The position of this point can be found with the aid of a small compass needle

Universe, as Einstein perceived it, made little direct progress, but certain side-issues were again seen to be important. The straight line was still the shortest distance between two points for someone (like Euclid) drawing on a flat piece of paper, but not to a motorist on the Earth globe, or to a traveller or observer in space. The geodesic, or 'straight' lines drawn in the hyperspace of four dimensions, are not straight, except perhaps to aliens from outer space. This is what Einstein's topology teaches, his geometry seeks to show space as it really is.

Part III

The Invisible Universe

Chapter 7

Smashing Atoms:
Looking at the Smithereens

There is an elegance in the way Nature operates at the quark level . . . it is important to understand the behaviour of quarks, for they are the deepest that we have managed to probe Nature's secrets.

Frank Close

Building on the past

Before we can understand what has been happening in twentieth-century cosmology we need to take a quick look at earlier work on the substances which make up the universe. Our current ideas are born of earlier ones.

This chapter deals with a central question of cosmology and explains how matter is organised. Understanding this means going into considerable detail, but this is essential to generate and illuminate general principles. The principles are what is really important, since they provide solutions to the main problems. The details are intended to explain only a few of the most general questions: Are atoms fundamental particles, and if not what are? How does the Sun work? Where does all the heat and light and other energy come from?

A tribute to Democritus

Earlier in this book, the Greeks were dismissed from consideration as begetters of Western culture. In mathematics, as well as in many other fields, their work is now known to be derivative from earlier civilisations such as the Egyptians, the Chinese and others. The Greeks transmitted selected parts of this heritage of thought, giving it a certain twist. Because of historical accident, other ancient civilisations have had to wait many centuries for the archaeologist's spade and twentieth-century scholarship to reveal their contributions, which were channelled as Greek gifts. However, it must be said that a few – a very few – did do really original work in science.

They included Thales, Democritus (his views were passed on), Archimedes, Epicurus (whose ideas were made known by the Roman poet Lucretius) and Eratosthenes.

Except for Archimedes, these men were not working scientists in our sense. At that time, there was no infrastructure (of interest, scholarship, personal association or involvement) to support the sciences in ancient Greece. The slavery and religious fundamentalism manifest there were death to the scientific method (and the converse is also true). However, these men were quite knowledgeable and asked the right questions, having had first-hand experience of the sciences developed in the East. Each had his own striking insight into the workings of Nature.

Most of the thinkers were atomists. This was the beginning of philosophical atomism, as distinct from the scientific variety which was delivered in Britain, trailing clouds of glory, more than two millennia later. The Greek thinkers suffered from a lack of tested empirical data, but were strong on reality-based observations. Democritus' views were the most highly developed. He taught that the universe was made of matter, and that matter was, in turn, assembled from very tiny particles which could not be further divided (the Greek word *atomos* means 'without a cut'). Thus material things, and this includes human beings, are built up of atoms.

According to Democritus, atoms are of many different kinds, just as things are. Things reflect or manifest a mingling or changing of the structures of the atoms of which they are composed. Things which occupy space, in other words all material bodies, are made of molecules (Democritus spoke of 'whorls'), which are produced by atoms colliding and joining together. These continuing collisions are the cause of change. Apart from atoms, Democritus said, there was only 'the void', that is, empty space.

Democritus also foreshadowed the laws of conservation of matter and energy, and made a remarkable and elaborate study of how we perceive the world through such senses as vision and hearing. Making allowance for differences in language, especially our extensive use of technical terms, his description is very close to the modern account provided by physics and experimental psychology. (In fact it was not until about 1850, with Fechner's law and the invention of a means of 'measurement', that psychology ventured into the area of psychophysics poineered by Democritus.)

Science and the atom

Science had made little headway since Democritus by the time of Galileo's arrival in the sixteenth century. Galileo and Newton gave us what looked, at least until 1900, like an almost perfect account of mass, force, velocity and acceleration. Newton accepted and developed Democritus' concept of the universe as being made of atoms. (Newtonians thought of them as similar to tiny but invisible billiard balls – hard and smooth, and subject to Newton's laws.)

The concept was also developed by Robert Boyle, who in 1661 wrote a seminal work entitled *The Skeptikal Chymist*. This book was the origin of modern chemistry and the death of alchemy. In it, alongside a critique of alchemy, Boyle provided us with the modern concept of an element. Elements are now defined as the simplest and most fundamental substances from which all material things in the universe are generated. Single elements combine with each other to form compounds. (Democritus was wrong in thinking that these were like the original atoms, as normally the compound differs markedly from the elements of which it is made up though it can be decomposed into them.) The number and nature of the elements, and the properties of their various compounds became the subject matter of the new science.

Boyle is also famous for revealing the relation between a volume of a gas, such as air (now known to be a mixture and not an element), and the pressure exerted on it. It contracts in proportion as the pressure is increased, and expands in proportion as the pressure decreases. The word 'gas' was coined by the Swiss physician Paracelsus (1493–1541) from the Greek word *chaos*, meaning 'swirl' or 'turbulence'. The Greeks knew that, although air was there (you could even sit on it in a gas cushion), its nature was entirely unknown. Under the influence of Boyle, gases soon became a prime subject of study. In the early nineteenth century they even became the subject of a parlour game used to divert the younger generation of ladies. The men would make 'laughing gas' (employed by barbers in pulling teeth) and the ladies would sniff it. It was rather like the more recent cocaine-sniffing parties but more educational. Gas became a popular topic at London's Royal Institution, which had earlier been set up by King Charles II (himself a strong supporter of science, pure and applied) for public lectures. At an evening class a popular lecturer would collect oxygen (discovered by Joseph Priestley in 1774, but anticipated by Scheele in Sweden) in glasses, using a pneumatic trough (a large glass basin filled with water), and burn brilliant tapers in it. He would then contrast it with other gases, which extinguished the tapers.

Faraday, at the time a bookbinder's apprentice, attended these evening lectures and sent his beautifully bound and illustrated notes, as a courtesy, to the lecturer, Sir Humphry Davey, who was also the Director. Davy, impressed, gave Faraday the job of washing the glasses used to collect the gases. In time, Faraday succeeded him as Director and became one of the greatest of England's scientists.

Concentration on gases as the subject of study was partly due to the fact that you could see how much you had, even of invisible gases. The covered, then upended, glass jars full of water received the bubbles of gas which displaced water from the glasses into the pneumatic trough. The scientists performing these experiments noticed an interesting result: that gases combine with each other by volume in very small number ratios (1s and 2s, or 2s and 3s). Knowing the formula of a compound, we can tell beforehand the relative volumes of gases and, by working back from their atomic weights, we can discover the relation between the weights of the elements in any compound. Many substances can be broken down into gases, the volumes of which are always very simply related to each other. For example, hydrogen and chlorine combine in equal volumes to yield two volumes of hydrochloric acid gas (the product dissolves and thus disappears from view, being soluble in water). Nitric acid, a liquid, can be broken down into three unit volumes of oxygen, one of hydrogen and one of nitrogen.

Similarly, it was discovered that solid compounds always consist of the same elements, and always in the same exact proportions by weight. For example, in table salt 23 grammes of sodium are united to exactly 35.5 grammes of chlorine. Also, two or more elements may combine in a variety of ways to give two or more compounds, for example, sulphuric (H_2SO_4) and sulphurous (H_2SO_3) acids, or phosphoric (H_3PO_4) and phosphorous (HPO_3) acids. In these compounds, the different weights of phosphorus, or of sulphur, which unite with unit weights of oxygen or hydrogen are related to each other in simple ways. These are the general laws of chemical combination. There is order among an almost infinite variety of substances or chemical compounds occurring throughout the universe.

When considering weights and volumes of substances combining with each other, they can best be explained by thinking of each compound as made up of different atoms which unite, each kind of atom having its own special weight. This is the logic which Dalton used to arrive at the atomic theory; it was based on the laws of chemical combination cited above. John Dalton was a Manchester schoolmaster and a Quaker. In 1808 he showed how Democritus' atoms explained these laws. In 1811, the Italian Amedeo Avogadro (1776–1856) developed the theory that volumes of gases combined, atom with atom, in small ratios by volume. (Put more exactly, he said that equal volumes of gases contain the same number of molecules. But this hypothesis was ignored for half a century.) The idea of atoms combining to form molecules was a powerful insight.

Dalton's work was the beginning of the chemical revolution in science. It was underpinned by the new chemical shorthand and by the 'grammar' and 'syntax' of chemical equations. The atomic theory was central to understanding the nature of the universe, enabling us to discover both the composition and speeds of heavenly bodies (from their spectra). It explained the enormous variety of substances and, more directly, provided the rationale for the manufacture of new substances. Faraday played an important part in this new industrial revolution, especially in the area of electrochemistry. He studied how soluble acids and salts dissolved in water, allowing it to conduct electricity. He established the laws of electrolysis about 70 years or so before J.J. Thomson and Ernest Rutherford experimented with protons and electrons. This early work explained how soluble acids and salts, in solution, broke up into 'ions' (charged atoms). In Greek myth Io was a female transformed into a cow, who 'wandered' the world to escape persecution by the goddess Hera. She is commemorated in the wandering ions, or charged atoms, which move to the positive and negative plates placed in a salt solution and connected to a battery. The ionic theory placed the general question of fission and fusion high on the agenda as a discussion topic.

The periodic table and atomic structure

The next momentous step was taken in 1869 by the Russian chemist Dmitry Ivanovich Mendeleyev. He verified that the elements were an organised system whose members could be classified and sorted by their atomic weights. The chemical and physical properties of the elements, he said, are a periodic function of their atomic weights. Earlier, many small-scale connections had been made between groups of elements, in 'triads' (threes) or 'octaves' (groups of seven) in order of their atomic weights. Mendelyeyev grouped them all in a single sequence, starting with hydrogen and finishing with uranium. The cycle repeated after every seventh item or element. (Inert gases, unknown till 30 years later, fell neatly into Group Zero.)

How was it possible in mid-nineteenth-century Siberia to discover the weight of the atom, something so minute as to be invisible (Mendeleyev was professor of chemistry at Tobolsk University). In those days, atomic weights were measured by comparing them to the imagined – because unknown – weight of the hydrogen atom. This unknown weight served as a unit, and other atomic weights were calculated, quite indirectly, from it. Between 1808 and 1869 the atomic weights of all the then known elements were established. Discovering how to do this was a noteworthy achievement of the scientific community, and it was Dalton in 1808 who led the way.

This is how it was done. Because hydrogen was the lightest substance known, its atom was probably the simplest and lightest. This, at least, was the hypothesis of choice, that is, a good place to begin. It was decided that its absolute (but unknown) weight would be chosen as the standard. By definition therefore, its atom was given a weight of one. Since the atom was simple, it probably combined with other atoms in the simplest ratios, that is, one to one (at any rate, we would assume so until we knew better). So other atoms would normally combine, by definition, in equivalent weights with hydrogen, *and with each other*. These equivalent weights for each element could easily be discovered by finding the weights of each which combined with one unit weight of hydrogen. In other words, we determine by an accurate chemical balance what weight actually combines with a standard weight of hydrogen or some other equivalent weight, such as eight units of oxygen. This is step one.

But, of course, the atoms may not combine in the ratio of one atom of hydrogen to one atom of the other element. To overcome this problem, we need to know the valency of the element (that is, its combining power). This will give us the ratio of the relative weight of the unknown atom to the unit weight of an atom of hydrogen. For example, take water. It is known from experiments (Faraday's method of electrolysis) that water breaks down to give us two volumes of hydrogen and one volume of oxygen. Hence, by Avogadro's principle (atom combining with atoms), the valency of oxygen must be two. If we *weigh* the two gases coming off at separate electrodes, we discover that one unit weight of hydrogen is combined with eight units of oxygen by weight. Now, since the valency of oxygen is two, the weight of the oxygen atom compared to the hydrogen atom must be 16 (2 for valency times 8 for the equivalent weight). The atomic weight of oxygen is therefore 16. (This method was later replaced by more sophisticated techniques. But its discovery, and Mendeleyev's work on it, was a breakthrough.)

By 1869, perhaps at many as 60 different elements had been identified and their atomic weights verified. Mendeleyev simply wrote down the names of these elements in the order of their atomic weights. He discovered that, after every seventh element, the sequence was repeated but on a different level. Similar elements were sited beneath one of the seven on the line above. The elements, arranged by weight, formed a matrix: the weights were aligned horizontally, the groups of elements vertically. It is a little like the octaves on a piano: seven full steps, then the notes repeat, but on a higher level.

In other words, the elements fall into what seem like prearranged slots, to yield about eleven-and-a-half octaves, each of seven elements (see the table on page 119). The spectrum of light seen in the rainbow is also arranged in a group of seven, as are musical sounds. Sounds are divided in octaves, where each octave has tones double the frequency of those in the lower octave. These similarities in the rules about sound, light and atomic structures strongly suggest a common ground-plan in the energy structure of the universe. We have almost found the master key; we may even do so by the end of this century. Together with the discoveries of the system function of black *and white* holes and of dark matter in the universe, this would be the crowning achievement of twentieth-century cosmology.

There are 92 naturally occurring elements plus 13 or so short-lived, elementary substances in the modern periodic table. Mendeleyev's table was fairly simple, but has now been made more complex to reflect up-to-date knowledge. Our deeper understanding of the table comes from awareness of the ever-increasing numbers of electrons in the orbital rings farther and farther from the nucleus. These in turn reflect the increasing number of protons and neutrons in the nucleus, the sum of which yields the atomic number. The sequence of elements, as well as the corresponding rings of electrons, are summarised in the table below.

The periodic system of elements

Hydrogen is an anomaly. It is usually given first in the table, being the simplest and lightest element, and sometimes shown at both ends, as we have done. There is a real rationale for this, based on valencies, but it is unnecessary to expand on it here. The formulae for the 92 elements can be shown in a table, thus:

Group 0 inert gases	Group 1 alkali metals		Group 2 bivalent metals		Group 3 boron group		Group 4 carbon group		Group 5 nitrogen group		Group 6 oxygen group	Group 7 halogen group
	a	b	a	b	a	b	a	b	a	b		Subgroups
	(H)											(H)
He	Li		Be		B		C		N		O	F
Ne	Na		Mg		Al		Si		P		S	Cl
Ar	K	Cu	Ca	Zn	Sc	Ti	Ge	Y	As	Cr	Se	Br
Kr	Rb	Ag	Sr	Cd	Y	Zr	Sn	Nb	Sb	Mo	Te	I
Xe	Cs	Au	Ba	Hg	La*	Hf	Pb	Ta	Bi	W	Po	
Rn	Fr		Ra		Ac		Th		Pa		U	

Group numbers correspond to valencies.

* A whole series of transition elements in the lanthanum subgroup has been omitted to save space, as they are not important here. The relationships between the elements are very complex and need a spreadsheet, or several tables, to explain them completely.

Maximum number of electrons in orbit in each ring

Very short period	1st ring	2 electrons,	elements H, He
First short period	2nd ring	8 electrons,	elements Li to Ne
Second short period	3rd ring	8 electrons,	elements Na to A

First long period	4th ring	18 electrons,	elements K to Kr
Second long period	5th ring	18 electrons,	elements Rb to Xe
First very long period	6th ring	32 electrons,	elements Cs to Rn
Second 'sharing' ring	7th ring	6 electrons,	elements Fr to U

We now know that it is the total number of electrons in the rings, and especially in the outer ring, which is vital. In particular an element's chemical properties and relations with other atoms are determined by two things: the nucleus (the number of protons and neutrons situated together); and the number of electrons in its rings (especially the outermost). This general principle is the one which decides the fine points in the material composition of the universe. The details need not be remembered; only understanding the basic principle is important.

Mendeleyev left three gaps in his table. He proposed that the empty spaces represented elements still to be discovered. However, he predicted the properties in detail of each of the unknown elements, giving estimates of their atomic weights, valencies, the types of compounds they would form, their physical appearance and a few other features. These elements were discovered over the next 17 years, and were called gallium for France (1875), scandium for Scandinavia (1879) and germanium for Germany (1886), after the countries where they were found. They had almost exactly the properties predicted by Mendeleyev.

Sherlock Holmes would have applauded, Watson would have been astounded. But Holmes would, as always, explain that there was no special mystery or magic in making deductions from established facts. Here too, there was no mystery. Mendeleyev took the four elements in the cells on either side of the missing element and, by averaging and allowing for trends in that part of the table, supplied the missing numbers to represent the observations.

From 1894 this effort was trumped by William Ramsay, who, working at Glasgow University, and prompted by the distinguished physicist William Strutt (Lord Rayleigh), discovered a whole group of rare atmospheric gases. They included helium (known previously only in the spectrum of the Sun), neon, argon, xenon, krypton and radon. These are now called 'inert' gases rather than 'noble', which was their first title. The group name was chosen to picture them as totally inactive, denying union to all other elements. (This is why they had remained undiscovered for so long.) In fact they do nothing except light up when a current is passed through them. The reason for this total inactivity is also clear: the outermost ring of electrons in each element is quite full (having eight electrons), so there is no room for any others to attach another element. If a new ring were to be started, the element would fall into another group altogether. Neon is the best-known of these elements because of its use in advertising signs. Since the Hindenburg disaster in 1937, when the hydrogen-filled airship caught fire at Lakehurst, New Jersey, helium, the other useful rare gas, has been used to fill passenger balloons in place of hydrogen.

Mendeleyev's prediction of the existence of other elements, and their subsequent discovery, as well as the isolation, one by one, of the seven inert gases, testifies that the order discovered by science simply reflects the order of the universe. Overall, hydrogen and helium have shown themselves to be the most common and important elements in the universe. They are almost certainly the first two steps in the building up of all the others. While at Cambridge, Fred Hoyle, a leader in British cosmology, worked out how this evolutionary process might work. Atoms of greater complexity are built up by fusion, beginning with hydrogen and then moving through helium to all the other elements. He took up the problem from the Russian–American cosmologist George Gamow, who was a pioneer in sponsoring much of the current emphasis on the 'initial singularity' theory – the creation story known as the 'Big Bang'. He detailed this in the 1930s.

Elements and their isotopes

We are now armed with the equipment needed to work out the nature, if not the origin, of the Sun. The simplest approach to this question is to outline the stages, or steps, towards the smashing of the nucleus. But to understand this, we first need to know about isotopes and their contribution to our everyday life.

In 1912 Sir J.J. Thomson discovered that if cathode rays are subjected to a magnetic field at right angles to their direction of travel, they are deflected from their straight-line path. The amount of deflection depends directly on the size of charge acting on the field and inversely on the weight of the particles involved. (We could have worked this out from general principles but Thomson's experiment is the verification.) The rays appear on the wall of the cathode-ray tube as a curve, in fact a parabola. The method provides us with a quick way to find out the different weights of the atoms of the elements. It is like a bathroom scale for atoms. The scale is etched on the cathode ray tube, so that you can see the deflection produced by standard atoms. Then you just read off the atomic weight of the unknown element, rather than having to perform a tedious series of routine analyses. Instead of cathode rays, you substitute a beam of atomic particles of the element you are studying, using the cathode-ray scale.

This all sounds very straightforward, but there were difficulties that had to be overcome in setting up and using such a scale. An anomaly was discovered when an element such as neon was tested. Neon's atomic

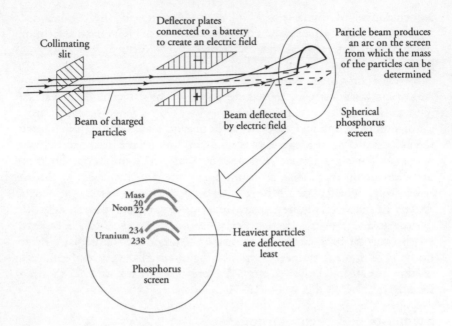

Figure 18   Cathode-ray tube used as a mass spectrometer

weight was known to be 20.2, but when this inert gas was tested on the cathode-ray scale, instead of one parabola fluorescing on the screen, there were two lines. The element tin produced more than 10 parabolas. The explanation is very simple, but it involved a breakthrough, and a new orientation to atomic structure.

In the case of neon, the simplest case, the positions of the parabolic curves on the scale indicated atomic weights of 20 and 22. It was clear that two kinds of neon particles (atoms) were hitting the tube. Like the alpha, beta and gamma radiation thrown off in the breakdown of an atom of uranium, the two kinds of particles, both identified as 'atoms' of neon, were sorted out by the magnetic field. In other words, neon was not a simple substance at all, but a mixture of two kinds of almost identical twin elements. They are chemically identical, but they differ very slightly in their atomic weights. (In fact they are like Siamese twins, being almost impossible to separate.)

What about other elements such as chlorine (atomic weight 35.5)? Maybe the fraction indicates that the element is really a mixture of two or more kinds of atom, perhaps half and half 35 and 36. In fact this is almost true: the atoms really weigh 35 and 37, with only a small amount of the 37 isotope in the mixture (this word comes from the Greek *iso* meaning the 'same', and *topos* meaning 'shape' or 'ambience'). We must conclude that all atoms are *discrete*, that is, the elements have only whole numbers for atomic weights. Those with fractional atomic weights are, in reality, mixtures of two or more almost identical kinds of atoms. This makes no difference to the chemistry of the element, and only the slightest difference to its physical properties. But it enables us to know atoms precisely as they are, with the correct nuclei and number of electrons.

In summary, the two elements both of which are neon, represented as Ne20 and Ne22, are almost identical, being isotopes. These two kinds of atom are mixed together, almost inseparably. The only real difference between them is that one atom is short of two neutrons in its nucleus when compared to the other. The nucleus of an atom consists of several heavy protons, each carrying a positive charge, and some heavy, but charge-free, neutrons. Each neutron is the same weight as a proton. (This, at least, was the view in the 1920s. It is now known that the neutron is very slightly heavier – an important factor in nuclear reactions.) The absence of charge, and a missing neutron or two from the nucleus, is responsible for the slight difference in the atomic weights of isotopes. The number of electrons disposed in rings around the nucleus must be the same in both types, because the charge on the nucleus is the same. This makes the chemical properties of the isotopes identical. This is what gave rise to the main problem of the wartime Manhattan Project to build an atomic bomb – how to separate the isotopes of uranium to select the one which balanced the equation and generated the planned atomic chain reaction.

The quark – smallest particle of matter?

The nonsense word 'quark' breaks with a long-established tradition of using Greek and Latin roots as names for processes and substances in science. Particle physics has been confounded by key words in American–English used in a sense different from their common meaning – words like 'up', 'down', 'color' and 'charm'. We need to accept these new terms (if not the spelling) to communicate about the new perceptions. In a fit of whimsy, the American physicist Murray Gell-Mann adopted the word 'quark' (originally coined by James Joyce in *Finnegans Wake*) as a technical term and defined it to explain the various kinds of subatomic particles that are liberated when the nucleus of an atom is destroyed by bombarding it with high-energy particles at high speed.

In passing, there is nothing special about Latin or Greek except that the words are there, and European languages have mostly developed from them, albeit in diverse ways. The adoption of Greek and Latin roots to make up the vocabulary of science was a stroke of genius. Scientists are most careful in choosing the precise word which enables them to specify the exact shade of meaning of each newly minted term. Regular academic conventions and standing international commissions make decisions about the vocabulary and syntax of scientific language. The use of name endings and classical nouns as nomenclature provides an international language understood by scientists everywhere, regardless of their native language.

Even when we come to revise our thinking about ultimate particles, later in this chapter, we can still accept all the analysis of matter reported so far; it has been tested and is valid. The terms 'element', 'atom' and 'molecule' still reflect an existing reality; the periodic system still stands as an inviolable contribution to our understanding of matter. However, as in our earlier remarks about quanta and relativity, we must now move to a deeper level, a new stratum of understanding, and abandon the idea that the atom is in reality the ultimate particle of matter. There is now a new level in our analysis of reality.

Targeting the nucleus

Comparing the work of the last quarter of the twentieth century with that of the Rutherford group in the first quarter, the recent study of sub-atomic particles looks like a rerun of Rutherford's research to discover the fine structure of the atom. Systematic bombardment of atoms, closing in on the nucleus and using various rays and high-speed particles as ammunition, was the prime technique in early atomic research. For example, in 1920 Rutherford fired alpha rays at atoms. (The nuclei of atoms were not affected by the most powerful magnetic fields then available, which accounts for Kapitza's interest in gargantuan electric and magnetic fields.) At the same time, alpha rays were 'scattered' by a thin piece of gold foil – and were even turned around to travel in the opposite direction. As Einstein had shown in the photo-electric effect (1905), this turn-around was most likely due to the alpha particle acting as a sort of 'bullet' and scoring a bull's-eye on the proton at the heart of a gold atom. In context, it was a clear demonstration of the 'strong force' which binds atoms together, making it virtually impossible to 'split' them by normal means. (This bombardment method, modified to generate a chain reaction, was later used to make the atom bomb.)

As Dirac had suggested of simpler cases, it was possible that many of the new particles (of which over 100 kinds have now been named) had a replica particle, identical in all respects to the original, but opposite in charge. His concept, that matter existed in two forms, as both matter and antimatter, has been borne out for even the smallest particles, called quarks.

There are three kinds of quark, so there are three kinds of anti-quark. Taking them together, the various combinations of these six tiny and invisible particles are now accepted as the core components of all protons and neutrons. As we have said, quarks have certain attributes which have been given bizarre names. First, there is a quality which sounds like dimensionality but isn't – quarks can be either 'up', or 'down', or 'strange'. Then they come in three 'colors' – red, green and blue – which refer to the polarity of the charges on the particle, whether positive, negative or zero. The charge is, of course, an electric charge, but unlike factory or domestic electricity, it does not flow, as it is bound in place. Thus, although the particle may be in very rapid motion (in which the charges share since they invest the particle), the charge is not free to move by itself, but has to be carried. But although the charge does not flow, it can be lifted and removed to another conductor. It can also be added to, and built up beyond the capacity of the particle to hold it. Then it flashes off like a miniature laboratory lightning storm. We can simulate all these effects, lightning included, by cranking the handle of a Wimshurst machine in the laboratory.

All the changes described above are electrostatic charges – hence the colloquialism 'static'. Static electricity consists of bound electrons or anti-electrons, imprisoned by the absence of any conducting material, such as a wire, for them to flow along. Such charges can be generated by friction, especially if the atmosphere is very dry. The charge may also build up spontaneously and spread over bodies, causing static 'cling' in dresses, it can make your hair stand on end, or bring a spark to your finger if you move it too close to the charge. When static electricity becomes excessive because of weather conditions, it produces the spectacular phenomena of thunder and lightning. Bodies of air rush against each other, driven by high winds to generate this form of electricity, by friction. Benjamin Franklin demonstrated these facts in a well-known, uniquely perilous experiment, conducting lightning to earth by a wire carried aloft by a kite.

In the case of quarks and atoms, it is the subnuclear work of 'colors' which holds the quarks together. Quarks can only exist as duos or trios, hence the emphasis on the sign of the charge. Forces of attraction and repulsion are generated between the different kinds of colours in different kinds of quarks. These colours resemble some natural processes in the macroworld. At the subnuclear level, quarks build up the 'strong' nuclear force which holds atoms together. Quarks also have a hypercharge which invests the whole particle, such as the proton (+), or electron (−), or neutron (zero). If the overall charge on the quark is positive, the anti-quark will be negative.

To be more specific, we first distinguish between two kinds of quark, remembering that they are compounded of two or three parts. One kind is known as a lepton (meaning 'light and insubstantial'), the other is known as a baryon (meaning 'heavy and robust'). The electron is a lepton, the proton a baryon. The electron is just under one two-thousandth of the weight of the proton.

Some quarks have another salient feature known as 'spin'. This is also described as positive, negative or zero, referring to clockwise, anticlockwise, or zero spin. The spin tells us what quarks would look like from various directions. Zero spin means that it looks the same from every direction, like a dot. A spin of plus one (+1) means it is like an arrow, and looks different from opposite directions, and so on. (There is no actual 'spin', as in skating. The terminology is purely hypothetical since quarks are not visible in this kind of detail.) Quarks are further distinguished by the existence or absence of mass, which is measured in electron volts. In these units, quarks may be relatively heavy, with a mass similar to that of a proton or neutron, but of course, much smaller than the nucleus of heavy atoms. Or they may be as light as electrons.

These attributes are important for Pauli's exclusion rule. This states that no two kinds of particle can exist together in the same state in the same atom (that is, with the same mass, velocity and position). This is a logical rule for thinking about *kinds*, or definitions, of particles; it does not refer to the ordered structure of particles such as protons, neutrons, and electrons which make up individual atoms. It is simply one of the rules of order.

Quarks are the ultimate 'stuff' of which the universe is composed. No single quark has been observed on its own – at least, not yet. They combine in pairs to give 'mesons', or in threes to give 'hadrons'. These are tiny particles, some of which may be long-lived. They have been known under their own names for a long time, for example as electrons and protons. But, in the main, quarks are quite transient, existing perhaps only for a few millionths, or even billionths of a second – long enough to be recorded by a detector device but hardly long enough for any significant analyses to be made.

Such evanescent particles can be discovered, and isolated, only by using giant accelerators. These operate on particles or atoms, causing them to strike a nucleus at very high speed, shattering the target. Particles, such as protons, gamma rays and heavy nuclei, are used to target various substances, breaking them up so that the fragments (fission products) can be examined. Various kinds of detectors exist to identify particles. In other experiments nuclei can be forced to unite to form fusion products. The more important particles are listed below.

Particle	Symbol	Mass	Charge	Spin	Lifetime	Force carried
Graviton	g	0	0	2	Infinite	Gravity
Photon	γ	0	0	1	Infinite	Electro-magnetic
Electron	ε	0.5	−1	1/2	Infinite	Electric
Proton	p	938 million	−1	1/2	Infinite	Electric
Neutron	ν	939 million	0	1/2	898 secs	

Estimated size of particles

Quark   less than the nucleus, are the smallest ever
Nucleus   one million-millionth of a centimetre (10^{-12} cm)
Atom   one ten-thousand-millionth of a centimetre (10^{-10} m)
Molecule   one thousand-millionth of a metre (10^{-9} m)

Mass is measured in electron volts, life in seconds. The table indicates how certain attributes are used to distinguish various types of particle. It includes only a brief selection of the best-known, long-lasting 'bosons', intended to illustrate the kinds of distinction made in particle physics. Particle physics is a leading edge in cosmological research, commanding massive resources of personnel and equipment.

High-energy studies are very relevant in cosmology, but the specific nature of the application still has to be clearly made out. This is because we know so little – indeed almost nothing which is not highly speculative – about the fundamental processes at work in the universe at either the micro or even the human level. Research workers in observatories see only the effects; research workers in laboratories discover facts about the nature of fundamental processes. But the two sets of findings have to be brought together. Maybe a greater diversity in the methods of setting up teams of workers is the prime need for the new century. The situation at the moment is somewhat similar to the period between the two world wars, when much energy was dissipated in ideological conflicts surrounding quasi-religious, metaphysical questions.

Essentially, we need to know how laboratory systems simulate what happens on a vast scale in the cosmos. The energy and mass transformations in the Milky Way and in outer space need to be simulated in high-energy laboratories, and the reactions related to cosmic events and processes. The velocities and the accelerations produced are similar in scale to those common in the universe, as shown in galactic and orbital redshifts and blueshifts. There are similarities between the laboratory fissions and fusions and the energy–matter–energy exchanges which create stars and galaxies.

The electron microscope – looking into the nucleus?

To describe a cosmic or microscopic object as clearly as possible, we need to be able to see it sharply enough to distinguish it from other objects in the background. For this, we need a source of illumination (usually a light source, but other kinds of radiation will serve), and a detector. Often this is just the naked human eye, but more commonly it is helped by a device such as a telescope, microscope, cloud chamber, bubble machine (to count small particles) or simply a piece of litmus paper.

In using an energy source to increase our powers of discrimination, we must ensure that the wavelength is smaller than the grain separating the structures being studied. The grain is reflected back in the image seen by the observer. To enable us to separate out the objects in the field of vision, the wavelength needs to be smaller than the junction between the different objects within the field of vision. This enables us to view our focal object and distinguish it from others forming its background.

The limit of our vision is given by this wavelength. It needs to be of the correct magnitude to single out and illuminate the smallest distance between the objects being viewed, and so enable us to 'see'. 'Seeing' is the power of discriminating. An optical device, such as a telescope, enables us to separate visually two objects about a thousandth of an inch across (about 0.025 mm; that is, about three times the size of a large molecule), or to see two close and tiny marks on paper which the naked eye sees only as one.

The electron microscope substitutes a beam of electrons for the beam of light. If we apply large voltages to the stream of electrons as it passes along the delivery tube, we can accelerate the electrons as a group, and so increase the resolving power of the microscope by decreasing the wavelength. In principle, this should work to an almost unlimited degree, depending only on the voltages available. If we use a special accelerator (a long hollow tube bent in a circle several miles in diameter), the speed of the electrons can be boosted many times by powerful magnets and currents. Scientists have even begun to see atom-sized particles. They are not bathed in light (photons), but in some other form of radiant energy. If we were to use accelerators several miles long, we should be able to focus on the nucleus of an atom, observe the protons and neutrons and, perhaps, even quark particles. Our accelerator just has to be long enough and our voltages powerful enough. The European Centre for Nuclear Research (CERN) in Geneva works on such problems in a network of national research centres.

Electrons (which travel freely in swarms, or circle as units in atomic orbits) are nearly at the limit of size and visibility. But quarks (bound duos or trios in atomic nuclei or in the many products of nuclear fission) would seem to be the ultimate, that is, the smallest particles of matter.

The secret of the Sun: nuclear power

To many people, the actual size, distances and composition of the universe, and how it evolved, are not a vital concern. Like ancient history, so long as scholars and specialists do their job properly, such things assume an interest only when some very remarkable discovery is reported. However, all this is rightly part of our education, helping understand the present, its processes and its outcomes, and as free citizens, we need this understanding.

We should, in particular, take an interest in the solar system, and in the processes centred in the Sun. To understand how the Sun 'works' is the key to many cosmological mysteries – including the source and survival of life on this planet (we know nothing about life on any other). We know that there are complementary processes of analysis and synthesis operating everywhere, both in the external world and in our minds. At the micro-level these reactions are called fusion and fission: one divides atomic nuclei (fission); the other, more frequent reaction creates complex particles (nuclei) by fusion. Such processes are found around the cosmos, producing enormous quantities of energy as a by-product of reactions. We have no control over these two cosmic activities but, by the end of the first third of the present century, we were able to duplicate some cosmic reactions by virtue of our knowledge of the basic mechanisms. We also have an idea of the prospects that they hold out for the human race – frightening or millennial, depending on political decisions made on our behalf.

Splitting the atom – nuclear fission

There are many ways in which we might have 'split the atom' or, to be exact, split the nucleus of the atom, just as there are many elements and isotopes that could have been tried. However, the principles can best be shown using uranium, especially the isotope of atomic weight 235 and atomic number 92. (It would be interesting to use the periodic system to check other possible ways that might work. The uranium route may be one of several possibilities.) The Manhattan Project to build the atomic bomb is a concrete example which worked. We will give the reaction in chemical shorthand, and in detail; the numbers are especially important, as the totals on both sides of the equation must balance. The numbers in the top line are the isotopic atomic weights, those in the bottom line are the atomic numbers, that is, the number of protons in each nucleus. The totals (236; 92) on the left side in the equation must be the same as the totals (236; 92) on the right side.

Uranium	+ Neutron	→	Barium	+ Krypton	+ TWO free Neutrons	
235	+ 1	=	137	+ 97	+ 2 times 1	= 236
U	Ne		Ba	Kr	2 Ne	
92	+ 0	=	56	+ 36	+ zero	= 92

The essential point is that, when we bring one neutron into close contact with one atom of uranium 235, they touch and react, producing two neutrons. The uranium atom splits into an atom of barium and one of krypton, but these are of no significance, being non-reactive by-products. However, if we have a mass of uranium 235, the new neutrons in turn react with two other uranium atoms to give neutrons, generating still more energy, and so on. In fact, we have started a chain reaction, where the next round involves eight neurons. It is like a dog chasing its own tail. The reaction continues to escalate until, within less than a second, all the U235 is exhausted. However, the 'critical mass' of uranium, that is the actual quantity of the pure element needed to continue the reaction to completion, has been carefully calculated. It also has to be held together long enough for the reaction to be able to be completed. This was the problem in making a bomb using the reaction. Once the reaction is established, more and more energy is generated – triggering an infinite series in which each number doubles the energy released in the previous one, as in the binary series:

1 2 4 8 16 32 64 128 256 . . . to infinity

The whole process is seemingly instantaneous. Vast quantities of energy are produced in a split second. The released energy is in the form of heat, light, sound and kinetic energy, with waves and waves of subatomic particles. This results in an explosion of stupendous ferocity.

Of even greater effect than this type of fission are the fusion reactions in the Sun, which produce quite unimaginable supplies of energy. The scale has moved from fission bombs based on uranium or plutonium to hydrogen bombs, where the principle is fusion not fission.

Nuclear fusion
There are two kinds of force acting in the everyday world – gravity and the electromagnetic forces. These affect large bodies, and assemblies of bodies, on Earth, in the oceans and in the skies. In the atomic dimension,

there are two other kinds of force at work. There is the 'strong' nuclear force, which holds the nucleus together. The fourth kind of force is unlike the others in that it does not work even at a moderate distance, but only when nuclear and subnuclear particles are very close together. One of the particles which 'carries' this 'weak' nuclear force is known as the 'gluon' (whose name perhaps gives some idea of this strange force, probably unique in physics).

When we talk about fusion, we refer to the union of the nuclei of two distinct substances brought together in a collision under conditions of extreme heat, such as at the Sun's centre (estimated at about two million degrees Celsius). There they fuse together, with the release of the most massive amounts of energy.

Where does nuclear energy come from?

When we speak of atomic energy we should really be talking about nuclear energy, since orbiting atomic electrons play no part in the reaction, except indirectly. Obviously their numbers and configurations will be profoundly changed by the transformations, but only the protons and neutrons in the nucleus are directly involved. We have described the fission and fusion reactions. However, the unanswered question remains, Where does the energy really come from? The chemical equations do not tell us, because they balance and show no additional energy on the right-hand side. The easiest way to resolve this difficulty is by a brief account of the context of discovery.

In 1934 Rutherford bombarded nitrogen with helium nuclei (alpha particles) derived from the disintegration of radium. The nitrogen atoms were converted to oxygen and hydrogen, in accordance with the equation:

$$N^{14} + He^4 = O^{16} + 2 H^1$$

(The numbers are atomic numbers, being the sum of protons and neutrons.)

The neutron was discovered in 1932 by James Chadwick, working in Rutherford's laboratory. This discovery speeded up progress in this area, breaking our dependence on alpha particles for the reaction. Neutrons are stable when in the nucleus of the atom, being tied down by protons. But outside the nucleus, as free agents, they are unstable and, when used as 'bullets', their decay can spark off nuclear reactions. So instead of the reaction quoted above, remarkable new possibilities opened up.

The vast amount of energy liberated in these reactions is due to changes in the atomic numbers, causing the *binding energy*, which holds the atoms together, to be redistributed and reconstitute new substances (elements). There is a vast reservoir of this binding strong nuclear force. Released in a bomb it causes enormous explosions as matter flies apart, enormous heat and colossal noise, and death and destruction to all living things. All the energy normally eked out by radioactive elements over millions of years is liberated instantly, with disastrous consequences. It is like having the Sun suddenly transported into your living-room.

Now, you may consider that the secret of the universe has been revealed. The details given in this chapter are clearly the concern only of chemists and physicists. But the processes are a matter for our common concern.

Chapter 8

A Cosmic Interlude

God does not play dice . . .

Albert Einstein
. . . nor is He impulsive (John McLeish)

Looking forward and back

A number of important issues are dealt with in this chapter that have so far been mentioned only in passing. They are summarised below and later discussed at length.

(1) One of the greatest advances in method was the first scientific classification of the stars made near the beginning of the century. It was the work of two astronomers toiling separately, and then uniting their forces. The first was a Danish amateur photographer–astronomer, the second an American professional astronomer. You might say that Hertzsprung and Russell gave the first rational (that is, scientific) classification of sky-objects in the heavens. This provided the life-histories of the stars and an account of their evolution. The authors' method was to cross-classify the stars by their brightness and spectrum type.

(2) Work with the Mount Wilson and Palomar mammoth telescopes led to the conviction, first declared by Herschel but widely recognised in the 1930s, that, when observing far-off stars or galaxies, we are also looking back in time at the history of the universe. This is a direct consequence of the fact, first established by Slipher in 1911, that the universe is expanding and that it takes many millions of years for light to cross the immense spaces of the universe. This seems to imply that the universe came into being at some specific moment in the past.

If we look at a star, the light which arrives at our point of observation must have left the star at the speed of light, that is, 180,000 miles (300,000 km) per second. It travelled all the way from star to telescope at the same speed. Looking into the far distance therefore means looking into the distant past – often the *very* distant past. The farther off the object in focus, the older it must be. The celestial events now visible through the telescope (such as a supernova or an eclipse) must have happened millions of light years before. Events now taking place in other galaxies will only be seen millions of years from now (assuming that there is someone around to see them).

By scaling down the size of our telescopes, or by refocusing them, moving by steps from far-off to near objects, we can work out the timeline of the universe in the same way geologists deciphered the history of the Earth. The principle stated, that the farthest-off are oldest, is an exact analogue of the geological law of superposition which states that in a succession of strata, the oldest rocks are at the bottom (unless there is evidence of inversion). Use of very large telescopes also gave cosmology access to the story of the universe, without having to make the physical efforts of the geologist.

(3) Jumping ahead from Hertzsprung and Russell, from 1910 to 1960, five years after Einstein's death, we find ourselves at a rebirth of relativity. A number of striking developments in observational astronomy, as well as several unplanned events, confirmed Einstein's model, bringing about a resurgence of interest in his work. Cosmology began to emerge from the shadow of subjectivism derived from Bohr and Heisenberg. Gravity became once again a key problem in the search for a unified field theory.

(4) The other question to be discussed is the nature of the invisible or dark matter in the universe. This is viewed in the context of the Big Bang theory. Was creation a single special act, as outlined in the Book of Genesis, or is matter being continuously destroyed and re-created at roughly the same pace, in a process which has had no particular beginning and has no foreseeable end? Is creation paralleled by a balanced, step-by-step destruction, a recycling of obsolete matter which reappears in other places? Is the universe a steady-state system where the amount of matter created over a period more or less balances the amount destroyed? This is how Nature operates on Earth in many other areas – in a series of natural cycles. There seems no reason to revise our model for the sake of a creation myth – elaborated in association with a misconceived view of planetary motions in a simple society. Maybe 'our' universe is not even unique; many others may exist out there, far beyond our ken. We are now well placed to collect information on these matters.

(5) As far as limits of size and visibility are concerned, electrons and quarks seem, at present, to be accepted as the ultimate constituents of matter. Electrons travel freely in swarms, or bound in atomic orbits; quarks travel as bound duos or trios in the nucleus of atoms, or are caught up in a multitude of particle species produced by fission. Electrons are made of quarks, which must therefore be even smaller than electrons

are. As far as particles are concerned, we need a simple and neat classification system based on a knowledge of *process*. This new theory seems to be lurking in the wings, waiting for a cue.

The Hertzsprung–Russell chart

Hertzsprung became interested in star spectra in 1902. He was looking for a connection between the brightness of a star (an indication of its temperature), and its spectrum. He figured that the red stars, which are cooler than others, would not radiate so much heat or light, and should therefore not be as bright as the much hotter blue and yellow ones. Star clusters, such as the Hyades, could be used in this kind of study. The stars in any cluster are all about the same distance from Earth, making relatively precise, direct measurement possible. These hypotheses held up except with Maury's C group, that is, exceptional stars with very dark, narrow spectral lines. Some of the red C stars were dim and cool, whereas more distant stars in the same class were very bright. These are now classified as red dwarfs and red giants.

An American, Henry Norris Russell, was busy with a similar star census, relating spectra to brightness. In 1910 he published the chart shown in Figure 19. He believed that this analysis revealed an evolutionary sequence. Over a very prolonged period, perhaps as many as 90 per cent of the main sequence would move out of this category into an adjacent one. Thus evolution is a continuing process but on a massive

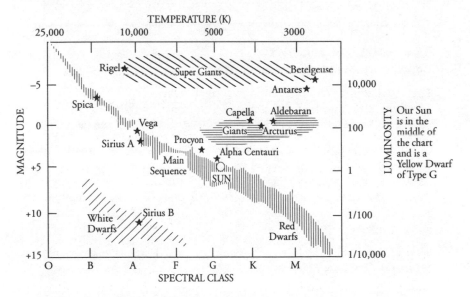

Figure 19 Hertzsprung–Russell diagram of the starry universe

time-scale. The classification was of a dynamic, but slowly changing universe. The most massive stars are very hot and very bright. They consist, in large part, of burning hydrogen. Having exhausted this fuel supply over millions of light years, they swell up to become giants and supergiants. Thus intermediate stars, such as the Sun, are now in the middle of the main group, or sequence. But, as soon as their hydrogen fuel is exhausted, millions of years from now, they will move into a new category as they change into giant stars. This phase accomplished, exhausted of fuel, they will survive as inactive dormant white dwarfs. The chart is, therefore, both a depository of stars and a record of the fossilised detritus of the universe. Russell did not solve the problem of stellar evolution. But he opened the question up and started the discussion. (Some typical stars, and their location in the chart, are shown in the table below.)

Typical Stars

Supergiants	Giants	Main Sequence	White Dwarfs	Red Dwarfs
Rigel	β Andromedae	Spica	Sirius β	Proxima
Antares	Aldebaran	Vega	Procyon β	Centauri
Betelgeuse	Mira	α Sirius	Eridani δ	Barnards' Star
μ Cephei	β Pegasi	Procyon α Centauri Sun	van Maanen's star	

The Greek letters α, β, δ and μ are used to identify single, prominent stars in the cluster named.

The rebirth of relativity

Einstein's three classic predictions were:

1) that relativity would explain the annual precession of the perihelion of Mercury (meaning, that the planet arrives consistently ahead of the time calculated in the position in its orbit closest to the sun; see page 97).

2) that a ray of light from a star on its way to Earth – say, the ray closest to the Sun, and so easily singled out during an eclipse – is attracted from its shortest path and, seemingly under the influence of gravity, appears to 'bend' towards massive bodies. (Newton made the same prediction, but Einstein doubled his estimate of the amount of displacement from a straight line.)

3) that light should show a redshift in its spectrum because of this gravitational influence.

The redshift was belatedly observed in 1960, but it was then agreed that this was not a definitive test of relativity, so second thoughts prevailed.

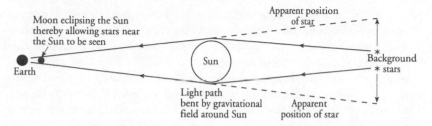

Figure 20   Gravity bending a ray of light

Right now, more than 40 theories of gravity are in the process of being tested for validity. No doubt this rather large sample of theories (which is an indication of our ignorance) will be reduced to one as tests continue to eliminate the non-starters.

Several events in the exploration of the universe prepared the way for the revival of relativity. As Max Planck once explained, scientific progress is not so much due to new discoveries, or even to the increasing wisdom of scientists, but rather, as in politics, it is achieved by the older members retiring or surrendering their power. We might call this one of the main lessons of history.

The first event signalling a new age was the flight of the Soviet satellite *Sputnik 1* on 4 October 1957. On board was a female Russian wolfhound named Laika. This flight was followed, at intervals, by nine more Sputniks, indicating that the Soviet Union had entered the Space Age armed with a programme of research. The ensuing race between the United States and the Soviet Union to conquer space led to massive expenditures which destroyed the Soviet economic and political system and, by the 'Star Wars' project, seriously affected the US economy.

A less spectacular event than *Sputnik 1* happened on 14 September 1959. This was the successful recording of a radar signal bounced back to Earth from Venus. Taken together, these two events indicated that the solar system was, so to speak, 'open for cosmic business' as a space laboratory. The launching of manned spacecraft as Earth satellites and space 'probes' made a reassessment of Einstein's theories an immediate, pressing problem for space travellers moving into completely unknown territories.

The four forces of nature

One of the most basic questions for relativity was the relation of the forces of the universe to each other. To Einstein, it was a matter of producing 'a unified field theory' for all forms of energy. Those who accepted the Big Bang explanation of the universe seemed to have no problem in agreeing that only a single force operated at 'zero hour', and that, as the universe cooled, this force which could act at a distance between bodies and in a straight line, split into four types. The Newtonian concept of force, involving bodies and charges in attractions and repulsions, had to be radically revised.

The new concept of 'force' fitted in with Minkowski's idea of a space–time continuum as a kind of warp in this manifold. (It is practically impossible to visualise this manifold in real terms. But, speaking loosely and figuratively, we can think of empty space as being in the form of a large bed or mattress. Massive bodies, such as the planets, affect this manifold much as a sleeper does the bed, leaving a large impression when he or she rises.) What we call a gravitational force in three dimensions becomes what must now be perceived, and thought about, as a depression in the space–time ('bed') manifold. What we see as a straight line has a 'dip' in the middle. We are back at the 'blind beetle' analogy (see page 104), where breaking into a new dimension (from 2-D representing a flat plane, to 3-D for a solid slice) alters our perception of reality. The straight line on a two-dimensional map becomes a curve (called a geodesic) on a three-dimensional globe. Similarly, our four-dimensional geometry must be able to describe a universe which is *pictured* differently from the early sketch we've become accustomed to.

Before discussing this concept, we have first to consider the four different types of force, how these relate to each other and how they act. It may be startling to learn that the first and weakest of these universal forces is, in fact, gravity. It is so obvious, universal and ubiquitous, that it seems it should be proportional to our expectations. Gravity was originally studied by an eccentric member of the British aristocracy, Henry Cavendish (1731–1810). Among his many notable discoveries was the element hydrogen, the most active and abundant substance in the universe. It plays a major role in cosmic affairs. Helium, a totally inactive gas and the second most abundant element in the universe, perhaps for this reason, was also first obtained by Cavendish, but not identified by him. (He isolated helium, but didn't recognise it as anything special, as it was mixed with all the other inert gases. He obtained these as a residual small 'bubble' by sparking ordinary air with hydrogen until both oxygen and nitrogen were used up. He wanted to discover whether anything would be left, and finished with a tiny bubble. But he didn't go on to explore what the bubble was made of, so failed to recognise the individual inert gases present, including helium.) However, he did discover the composition of air and the fact that if hydrogen was burned in it this produced water. This was another blow to the Greek theory of the four 'elements', since clearly neither water nor air were elements in the Greek sense. The third of the four, Earth, was also obviously no element, but a gross mixture of many different compounds. Finally, fire, the fourth, was not a 'substance', so it did not combine or 'mix' with anything, as elements should.

Cavendish made many other scientific discoveries. For example, he invented the torsion balance in 1798, to measure the force of gravity, checking the pull of the Earth by setting it against a large solid body in his laboratory. He also measured the same force, using a plumb line, acting between two solid bodies in the same room. In the mid-1800s the Hungarian Baron Eotvos perfected this torsion balance to make it the most precise instrument available for measuring mass (estimated as weight).

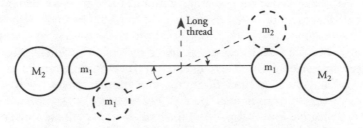

Figure 21   Torsion balance 'weighing' the Earth
The balance has 2 large weights (M_2) set up on either side. The balance is set swinging and the period of oscillation is estimated as precisely as possible. From this, the value of g (gravitational constant) can be found and thence the mass of the Earth calculated. From Newton's gravitational laws the Earth's mass is given by $Me=gR_e^2/G$ where g is the acceleration of a body due to Earth's gravity and has a value of 32 feet per second every second, and R_e is the Earth's radius (about 8000 miles).

Cavendish used the torsion balance to prove one of the most basic principles of Newton, and also a basic assumption of relativity, that inertial mass (the mass that resists movement) was the same as gravitational mass (which affects and controls the movement of bodies). This means that the resting mass of a body is the same as the mass of the body when falling freely. In other words, gravity was the single force which affected all falling bodies as well as bodies at rest, or bodies moving freely as planets do. Newton is supposed to have had this insight while sitting in his garden – hence the myth of Newton's apple (see figure 22). The truth is that Newton showed mathematically that the same force which explained the fall of an apple also accounted for the motion of the planets and other celestial bodies. (As far as I remember, he didn't even mention apples!) Galileo had made the same judgement a century before Newton, but he tends to be downplayed in Anglophile accounts of the story of physics.

Figure 22   Newton's 'apple' and planetary motions
Using his law of inverse squares, Newton calculated that the moon is
subject to the same attraction as bodies in free fall on Earth. The strength
of the force of gravity on a body is determined by the body's mass and its
distance from the Earth's centre. Gravity holds the moon in its orbit about
the Earth and causes falling bodies to accelerate at 32 feet per second every
second.

The second of the four forces which make up the unified field theory is
the electromagnetic force. This governs the flow of electrons in current
electricity, and the behaviour of electric and magnetic fields. It is many
trillions of times stronger than gravity. For example, a tiny magnet can
support a massive iron nail in opposition to the gravitational force of all
the planets working together (Earth being closest and most significant)
and the resultant (sum of) force of all the matter in the universe.

The third force acts in atoms holding the positive protons and the
uncharged neutrons in place. It is very powerful and is known as the
strong nuclear force. The positive and negative charges in the atom
exactly balance and, as a result, the atom itself is neutral. But the strong
force makes it practically impossible (but not quite!) to separate the
components of the nucleus.

Lastly, there is also a weak nuclear force which acts at very short
distances indeed and not at all at any greater distances. It is about 10^{12} (10
million million) times weaker than the strong nuclear force. It controls
the union of quarks, such as those making up neutrons, protons or
electrons. This force causes some radioactives decay (beta decay).

Curved space–time
The 'Yellow Submarine' in the well-known Beatles song is Liverpool
slang for a madhouse, or more politely, asylum. It differs from a mental
hospital where therapy is provided; its main concerns are shelter and
security. Maybe one of the group had been using a relativity text as light
reading, and the idea of gravity as warped space–time inspired the song.

As we said earlier, Euclidean geometry is an ideal system for architects,
surveyors, temple builders and anyone not interested in anything outside
city limits. This was true of the ancients, Egyptians, Greeks and many
others. Euclid's geometry is still valid for small, flat surfaces such as maps,
but only as sketches for short journeys by road or sea. It was Magellan

who first created problems for map-makers by sailing around the world. It became a practical necessity to accommodate maps to the fact that we live on a sphere and not in a level, flat field. No matter which 'projection' of a sphere on a two-dimensional surface was used, printers of accurate maps discovered that they had to commit extreme violence on vast areas of the three-dimensioned Earth to represent it on a two-dimensional map. Continents had to be bent out of shape (and still are); stretches of territory on the map looked to be running north and south until you studied the printed 'legend' (a well-chosen word indeed) at the side.

Mercator next became the name to conjure with. This was because, in his 'projection' of the world sphere onto a square surface, he spread out the inevitable errors, siphoning them off to high latitudes and desert places, and averaging out the remaining error by spreading it around in the 'advanced' countries where his maps would be in greatest use. Having learned to accommodate our ideas of space and time to living on a globe, we were projected into the space age with amazing newsreel footage of astronauts in space.

The concept of weightlessness signalled the end of trying to understand for many people. By giving up at this point they were excused from the need to grapple with the core of Einstein's theory, namely, that the universe is a curved space–time continuum. (If we dismiss everything we don't understand as so much 'yellow submarine' talk, we can still survive in a consumer group even if, at times, it may remind us of living in a cuckoo's nest.) Weightlessness is a matter of 'being in the wrong place at the right time' (that is, becoming a kind of mindless pawn at a 'neutral point' in a gravitational field). In the astronauts' case, it means that they have entered a region where the gravitational forces exerted by all the planets and other masses, acting as they do in all directions, are in balance and so cancel each other out.

The 'field' concept of gravity

The astronaut in weightless conditions becomes like a very tiny magnet, say the needle-pointer of a small pocket compass. We can place a large bar magnet on a piece of drawing paper and move the compass from one end to follow the changing pointer. Assuming that we mark the small changes of the needle as it aligns itself with the line of force, the compass orients itself more and more in the direction of the Earth's magnetic field or force. In this way, we can map a horizontal slice of the Earth's magnetic field. Placing the bar magnet with its north-seeking pole facing south, in opposition to the Earth's magnetic field, we can map the field of force as before. This time we discover two points, one near the north pole of the magnet and the other the same distance from the south end, where the Earth's magnetic field is exactly balanced by the large bar magnet. At these two points the compass needle swings wildly at first, then settles down but in a random direction. These points are the two neutral points, where the all-pervasive magnetic field of the Earth is exactly matched and neutralised by the contrary field of the magnet.

We refer to 'fields' of force, but these are actually in three dimensions, stretching up–down, left–right and front–rear. The 'field' is really a 'sphere' of force, the force becoming weaker and weaker as the distance from our laboratory bar magnet increases. The Earth behaves as if a huge magnet were buried in it, running almost due north and south. The Earth magnet creates a huge magnetic field, and it doesn't matter much where you are, the magnetic force is about the same.

Einstein thought of the Earth's gravity field, in common with other gravity fields, as the effect of the flow of gravitons, just as sunlight is supported and transmitted by a stream of photons. These travel at the speed of light.

Energy particles and waves: light and gravity

These ray–particles (gravitons and photons) might be pictured in the image of the old-time commercial traveller (the stereotype caught up with the farmer's daughter). Not having any independent being while travelling on the train, he materialises as a person only when he steps down from the railway car. On leaving the train, he casts aside his Traveller's Mode as his feet hit the ground, and assumes his Commercial Role to call on his regulars. Still in this material mode, he performs other 'commercial', and perhaps some leisure activities which mark him out as a human being. He then 'entrains' and immediately resumes his role as 'traveller'.

The graviton has the same kind of duality. Like other forms of energy, in certain situations and depending on context, it is constrained to act as a particle. In others, it switches, to travel as a wave. The magnet, like light and gravity, is a source of another kind of non-material energy. This is produced by the alignment of the small particles that make up the body of the magnet, being themselves tiny magnets. They are aligned in parallel rows in the iron ore (magnetite), which magnets were made of before Faraday discovered how to make them by electrical induction. Before Faraday, you stroked a soft iron rod repeatedly with the ore, in one direction only, along its length. It was thought that, as a result of the stroking, the tiny magnets in the bar, arranged randomly, were attracted by the magnet, and aligned by it to point all in the same direction, one end to north, the other to the magnet's south pole.

Dark matter and black holes

At the centre of our thoughts about the universe is a picture of the starry heavens. This holds our attention, channels our thinking about the subject and dominates all our talk. This is the universe reflected in all our reading. More than 90 per cent of the technical writing about it refers to the incredible events taking place before our very eyes, or recorded in previous writings. Literature, when it looks for attention-capturing figures of speech, needs no other images besides stars, planets, galaxies and other visible (because illuminated) objects. Our reading conditions our thinking. Even when we begin to use our ears too, trying to hear 'the music of the spheres', we don't really listen because we lack any training in the art. We don't feel happy until we can attach sounds to an identifiable, and therefore visible, source. Either the Sun, or a radio star, or a galaxy – some visible form – then becomes the focus of our attention. We find it too dull to contemplate the notion that the invisible, but very active, energy universe must figure in our explanations and theories about the origin and the continuing life of the universe. We want to know *how it looks*.

It is a shock to learn that the distribution of light from the almost infinite number of stars, galaxies and other celestial bodies reaching our telescopes is nothing like a register of the mass distributed throughout the universe. Some authorities have estimated that more than 90 per cent of the mass of the universe is 'dark matter', that is, invisible, therefore unseen by observers. This does not mean it is dead, or otherwise inactive, non-functional. On the contrary. It acts in similar fashion to the human skeleton. This holds the pieces together, in place; it gives structure to the organism. It enables us to act on our surroundings by providing leverage, 'body' and 'direction' to our existence. Unseen matter behaves in exactly the same ways as our non-visible skeleton; it is the unseen body of the universe. The analogy between the universe and the human organism has been drawn many times by scholars, but they haven't taken it far enough.

Among the massive collection of nebulas and galaxies sprinkled throughout the universe, many are rotating at great speed around a central core. (You may recall the Kant–Laplace nebular hypothesis, page 56.) Certain galaxies resemble the firework called Saint Catherine's wheel. This is nailed to a door or wall and set in violent circular motion by a continuous jet of exhaust gas from the explosive. Some galaxies look exactly like this wheel, but obviously don't move so fast. The danger still exists that the circular motion, especially if the galaxy is disc-shaped, will cause massive bits to be thrown off by the momentum resulting from the rotation. By analysing the two spectra of such saucer-shaped rotating galaxies, we can determine the speed of their rotation, from the redshift of the approaching 'half', confirmed by the blueshift of the retreating 'half' of the revolving galaxy. Many galaxies exceed the limit of velocity needed to maintain cohesion. In normal circumstances this would lead to their break-up. The fact that it doesn't, could well be due to the 'braking' work (gravitational attraction) of the solid, invisible matter spread throughout the body of the galaxy. 'Empty space' is probably full of the stuff; it certainly contains a vast quantity of cosmic dust and vapour, which are, of course, material in nature. We see it only if it blocks the light from some celestial body.

But the really important masses of invisible matter are probably those constantly attracted, and swallowed without trace, by the so-called 'black holes'. These holes are scattered throughout the universe as last relics of the novae and supernovae which flash out every so often and light up the whole sky, sometimes for very lengthy periods. They then burn out to become invisible. Black holes very vigorously attract any matter that is close enough. It disappears over the edge, without trace.

In fact the force is so great that nothing, not even photons, can escape from it. So, over the edge, or even looking in, the hole must be totally opaque, as no light can escape.

This is, of course, almost pure speculation. The only saving grace is that black holes do seem to exist, and probably in very great numbers. As we have indicated, they seem to be the final stage of the super-bright stars (novae) which flare up about once a year in a galaxy. This must mean there are millions of them scattered around, at least one per galaxy. Speculating wildly on their function (which is all anyone can do at present), we may suggest that, once in the hole, some natural process may cause solid objects to disintegrate by collisions, maybe just below the speed of light, prior to being converted to energy.

After such disintegration, the recycled energy can then be released, by overall diffusion, or possibly be expelled, through a special exit, or by some unidentified process, back to the mainstream of observable events. Some scientists even speculate on a more dignified exit through the invisible end of the black hole into a new dimension (the fifth) of existence. The Oxford mathematician Roger Penrose suggested that inside every black hole there is a mechanism (a 'singularity') where collapsing matter, or any other detritus, is crushed out of existence. Or the matter might pass through a 'worm-hole' (a structure that enables materials to scrape through a forbidden boundary) into another space and time. These so-called 'white holes' may provide an exit for the energy generated in the annihilation of matter. This energy, according to the steady-state model we are describing, fountains out as showers of photons, radio signals and in other forms.

This model seems to present a possible alternative to the singularity creation theory. One possibility is that the black hole is just one unit of a gigantic sewer system, the *cloaca maxima* of the universe. Here the digested residues of exploded stars, aborted galaxies, cosmic dust and other remanent or surviving matter are recycled as in a treatment plant, as part of a continuing cycle. The recycled mass, converted to energy, could well be the source of the 'cosmic hum' attributed to the 'original singularity'; that hum need not be the microwave residue from the original Big Bang, as is claimed by those who accept this view. Some even talk about being in a position to 'read' God's mind in speculating about these matters. When Saint Augustine was once asked, 'What was God doing before the Creation?', he replied 'Preparing Hell for those who might ask such a question'. An alternative answer might be 'Working out the energy and mass change-overs needed for a waste-disposal plant in harmony with Our principle of economy and Our other recycling triumphs.'

We do not need the wizardry of the calculus to find out what happened in the first millionth of a second after the tiny cosmic egg cracked. The Russian cosmologist Lev Davidovitch Landau expressed the thought in 1932 that, in the timetables of the kind referred to, 'assumptions are made only for mathematical purposes' (usually to solve a particularly nasty equation) 'and have nothing to do with reality'. The trouble is that some mathematicians tend to forget why the assumption was made in the first place and when to stop making it.

Creation – a 'Big Bang', or just a continuous hiss?

It is now the orthodox view that the universe had a definite beginning. This is not just a bare assertion, but is supported by a great deal of circumstantial detail. In fact, the exponents of this theory have given a blow-by-blow account of what happened in the first 6 or 10 seconds of the universe. This is in the form of a breakdown, in some places in millionths of a second, especially at the beginning of this singular event. Such enlightenment about the event (assuming that it happened) is obtained by applying advanced mathematical procedures to a small number of equations. Instead of the word 'Creation', we are invited to use the word 'singularity', because it only happened once, and under very special conditions, and can never happen again. 'Creation' is not a 'good' word.

Since we can have no first-hand experience of this, or any similar event, I see no advantage in making preternatural calculations. I shall therefore describe the total process without bothering about the calculated times at which the events were supposed to take place. Far from adding 'verisimilitude to an otherwise unconvincing narrative', in the words of W.S. Gilbert, the main effect of the detailed timetable is to render the whole story preposterous. I shall stress only rational details.

In the beginning there was the void, mere nothingness, stretching to infinity in all directions. It was empty apart from a single point which was so small as to be invisible. Then, for some unknown and uncaused cause, this point started to expand. In a fashion and by a process not explained, it already contained all the matter now in the universe. The hatching of this primeval egg caused a massive explosion, accompanied somehow by incalculable temperatures, billions of degrees Celsius. However, with the void being at a temperature of Absolute Zero ($-273°$ C), and infinite in extent, it did not take long for the new universe to cool sufficiently for the hydrogen atom, the simplest form of matter, to emerge and become established. (No atoms could exist before, as the hellish temperatures stripped off any electrons which might form and circle nuclei.)

Other elements appeared, one by one. The progression started with hydrogen, and continued through the periodic table as far as helium. The process involved was fusion, nuclei coming together and fusing at enormous temperatures. In due course, it produced the heavy metals such as gold and silver, and finished with uranium. This process had to wait for the temperature to drop, because not even the elements could survive the initial conditions of the universe. Thus, the stars and galaxies, the solar system, Earth and, eventually, life were provided with a suitable environment, and appeared.

In 1965 two Americans, Penzias and Wilson, working for the national telephone service, accidentally discovered a microwave background radiation. Signals from all parts of the universe transmitted this radiation at the same wavelength, and indicated a common temperature of 2.7 degrees above Absolute Zero throughout. Although at first Penzias and Wilson seemed unaware of the implications of their discovery, it was believed to confirm a conjecture by George Gamow in the 1930s that there should be just such a microwave radiation which had continued from the beginning and might now be capable of being monitored. Putting two and two together, Penzias and Wilson decided that they had verified that a single event had given rise to the universe. This is the theory, set out in the 1930s by George Gamow, that Hoyle jokingly referred to in a broadcast as the 'Big Bang'. Fred Hoyle had also presented an alternative view, which he named the 'steady state' theory. Although dismissed (temporarily, I am sure) by contemporary opinion, it has more the look of a purely scientific theory uncontaminated by apologetics. It may be 'reading the mind of God', but at a respectable distance, at second remove, so to speak.

Let us consider for a moment the rather convincing arguments for the theory that the universe was created in a single mega-explosion many millions of years ago. There is evidence that the universe has been expanding over many millions of years. This was shown by Slipher, and confirmed by Hubble. Observations confirmed Hubble's Law: that the galaxies are receding from each other, the more distant the faster the recession. The heat energy generated at the original singular 'creation' explosion was essential, being converted into the dynamic energy needed for the expansion of the super-colossal mass of material. The residual heat was sufficient to raise the temperature of matter to billions of degrees. This period lasted an estimated 6 seconds. The date was about 10 thousand million years ago, that is 10^{10}, or maybe it was more. (There is more than a slight degree of probable error in this estimate.) The laws of Nature did not yet exist, only one of the four basic forces of the universe did. Like the story of the Creation in the Book of Genesis, believing this account requires a leap of faith, indeed several leaps. There is, and can be, no direct, specific, unequivocal evidence for this event. It relies on the credibility of the account. This credibility, in turn, depends on not contradicting established facts. There is also the related but independent question of plausibility. In other words, does the theory blend harmoniously into the general picture of science as this has developed over the centuries? Or is it mostly based on ideology (that is, is it similar to Nazism, communism, fundamentalism, or any other uncritical faith replacing a reasoned belief in process)? What we are talking about is a basic non-commitment to the truth about reality. As Tertullian once said, *certum est quia impossibile est*, 'it is certain, because it is impossible'.

The best supporting evidence for 'the singularity' is that energy distribution in the universe indicates a mean temperature of + 2.7° Kelvin (about three degrees above Absolute Zero). This has been associated with the microwave 'background' radiation which comes from all directions – a 'hum' if you like, received *at a particular radio wavelength* at all research centres. The other seemingly undeniable fact, perhaps related to this, is the ongoing expansion of the universe. It is difficult to explain how this could come about other than by this favoured theory. There may be a simpler explanation, but it perhaps significant that it was a Belgian priest–astronomer, Father Lemaître, who first proposed the theory in 1926. His views were soon afterwards declared by the Pope to be compatible with religion and revelation.

Chapter 9

Energy Sources in the Sky

I'll put a girdle round about the earth in forty minutes.
                    Puck, *in* A Midsummer Night's Dream, 2, i, 170

Ariel (sotto voce)*: 'I'll do it in a hundredth of a second . . . Try me. Just blink.'*

                                        (adds John McLeish)

Setting the scene: a personal memory

My father was what was called a radio ham. He didn't set up a transmitter, but made radio receivers for his own use, and for friends, as a hobby. When he began on this, I had not yet started school, but I became his little helper. I was quite fascinated by the new toys – twisted bits of wire, simple tools like wire-cutters, strippers and nippers, not to mention such 'parts' as cats' whiskers, wet (acid) batteries, which needed recharging about once a month, variable condensers, and later, valves. All of these, and circuit diagrams, could be purchased for pennies from a large 'wireless' shop nearby. As I grew older, we talked about the pioneers of radio, Marconi (died 1937) and Oliver Lodge (died 1940), whose activities were reported in the amateur radio magazines, which my father explained. That way, I at least learned the names of things. We both became very excited during the build-up for that day in 1924 when the BBC started its national broadcasts from the Crystal Palace in London. I can still hear in my mind's ear the first announcement on public broadcasting in Britain. My father and I immediately became wireless addicts. There was only one problem. Our crystal set selected stations by means of a thin pointed wire, called the 'cats' whisker', which was mounted on a 'crystal' of germanium to move easily to contact different areas. The crystal enabled us to 'tap in' to radio signals picked up by the antenna and listen to programmes from all over the world – including Dublin, Berlin, Paris, Moscow and London (soon Kelvingrove too, the Scottish station). We listened to all of these, somewhat awe-struck by our secret connection with these capital cities, and by our own temerity. But there

148

was an annoying constant hiss – my first introduction to the cosmos. It altered in volume without any reason, so far as we could tell (we listened through earphones). There were also irregular fadings and unpredictable volume changes in programmes. Sometimes strange noises and voices broke through, speaking in foreign tongues, and there were scraps of music. Crackles and other noises sometimes took over for brief periods, drowning out reception. We considered these inevitable, since hardly anything was known about what happened between the broadcast station and the receiver. We knew about the 'Heaviside layer', which bounced 'wireless waves' back to Earth, and we believed that it was transient 'holes' in this layer which allowed programmes to escape to the outer universe and which let in 'foreign matter' (called 'static') to annoy or fascinate the listener.

Over the years, valves (known as 'tubes' in the United States), became available, and later solid-state electronic components (transistors) replaced germanium crystals, and most other parts as well. Reception constantly improved as a result of research by Phillips and others, and loudspeakers took the place of headphones. At first, acid-lead batteries as well as large dry batteries had been used to power the system, and there were also small dry batteries to 'bias' the grid (in those days domestic electricity was not common). These were all interconnected by special wire and serviced by the amateur 'radio hound'. Over time, a whole subculture and language grew up to describe installations such as ours. The adjective 'Heath Robinson' was invented, after the artist who drew fantasy cartoons for Punch mocking the impossible contraptions (some much worse than ours!) made by committed amateurs.

As new materials and components came in, alas, our old-fashioned 'ham' days were numbered. The whole arena deteriorated as commercial exploitation took over.

The invisible world of radio waves

A similar deterioration was noted as the telephone conquered the mass market, beginning in the United States. The Bell Company, always a prime mover in the field of science, decided that to discover and eliminate the source of extraneous noises would vastly improve the system to the benefit of subscribers. These noises were like the hissings and cracklings known to radio buffs as 'atmospherics'.

The opening of research into these secret cosmic 'signals' started in Country and Western style in a potato field in New Jersey. Karl Jansky, a professional radio engineer, was employed to find out what caused all this interference 'on the line'. He succeeded in getting rid of most of the weird noises, but the persistent hiss remained. In 1928 he ventured on a new tack by dismantling an old Model T Ford so that the chassis could be pushed on rails across a field. It carried a Jansky Heath Robinson contraption, consisting of 400 feet of plumber's brass pipe fitted together to make an antenna (or aerial), like those used in radio reception but more elaborate. The antenna was carried on a turntable which ran on a circular track, and was connected to a listening device inside the house. (Jansky worked at home because of ill-health and used the kind of 'ham' radio system we had.) The antenna revolved on its own axis at a steady speed, about three revolutions a minute. It could be directed at any sky object, such as the Sun, for long periods, or it could simply be set up and left to rotate with the Earth's motion surveying a swathe of sky. Next day, it could easily be moved to a new position to survey more sky. This was a very ingenious, if primitive and quite inexpensive, piece of scientific apparatus.

Using this apparatus, Jansky quickly identified the hissing noises as the same as those which interfered with radio programmes. He noticed that the 'atmospherics' varied in intensity with changes in the patterns of sunspots. This proved that the hiss on the telephone line (which sounded rather like trapped air being keyed out of an old-fashioned hot-water radiator system) was associated with the Sun. Jansky showed this by directing his antenna towards different areas of the sky, including the Sun. The hiss varied in volume according to the time of day, but reached a maximum at a particular time. This time also varied, the maximum hiss appearing four minutes earlier every day. This discovery implied that the hiss was generated by emanations from heavenly bodies, and particularly from the Sun.

Jansky deduced this conclusion from the fact that the invariant four-minute interval each day corresponded exactly to the difference between sidereal (star) time and solar time. The Earth travels round the Sun in approximately 365 days and a quarter. The Earth also rotates on its own axis (the imaginary line joining the two poles) once every 24 hours. This rotation causes the succession of night and day, as the Earth turns away from the Sun's rays and back again. The two systems (solar and sidereal time) do not quite agree about the length of the day. In solar time it is set by convention at 24 hours, but in sidereal time (which takes the fixed stars, not the Sun, as reference), the day is equivalent to only 23 hours and 56 minutes of solar time. So if we measure each day in sidereal time, the hiss would be maximal at the same time every day in sidereal time but would differ by four minutes in solar time. At least, this seemed to be a sound initial hypothesis. It took Jansky four years of hard work to establish it, but it was a mighty achievement.

Jansky's report of his findings created some degree of popular interest when it appeared in the newspapers, but little professional reaction when it was published in an obscure technical journal. It did, however, inspire the radio engineer Grote Reber, in Wheaton, Illinois, to investigate the subject for himself. Reber designed and constructed a radio disc (as they were then called) in his garden at a cost of $2,000. (This was during the Depression. The sum was equivalent to £400 at that time – just less than two years' salary for a schoolteacher.) The disc was bowl-shaped, over 30 feet (10 metres) wide, and made of copper wire on a rigid frame. Reber made a habit of going to bed early, rising at midnight and working at his self-assigned task until six o'clock each morning, when he set off for work.

Reber confirmed Jansky's view that there were other sources of radio-length waves besides the Sun. One source he found was in the Milky Way, with its centre somewhere in the constellation Sagittarius. He also identified other sources, notably in the constellations Canis major and Cyngus. The radio signals (they are called 'signals' but are really just noises) often came from what seemed bare patches of sky with no visible stars. This, of course, was a very significant finding.

Jansky and Reber, between them, established the new science of radio astronomy, and devised the essential instrument for developing it: the disc antenna. But nobody paid much attention until after 1945. Reber did receive some money from a research fund, but the intervention of the Second World War seemed a good enough reason at the time to avoid work in what appeared to be rather a narrow field.

In the meantime radar had been developed and accepted by the British government as being of national importance. Radar is the system where radio signals are directed at an invisible target which, when found, reflects most of the energy back to the source (in popular language, the signal 'bounces'). Since radio waves travel at the speed of light (300 metres or 1,000 feet per second) the distance of the invisible object can readily be determined. Radar was first developed in the mid-1930s as a result of the work of Sir Robert Watson-Watt. It became a national priority under Churchill in wartime Britain. It was used to identify approaching bombers, naval vessels and other large objects, and was linked to an early-warning system. In 1945 this equipment became surplus to Army requirements, and most of it was handed over for scientific research.

The fiasco of 12 February 1942, when the warships *Scharnhorst* and *Gneisenau* passed undetected through the English Channel, drew attention to the fact that radar signals could be 'jammed' (as Jansky had shown) by the unexpected activity of galactic radio sources. But the jamming occurred only during daylight, especially if the radar aerial was pointing directly at the Sun.

This was really the beginning, or 'quickening', of radio astronomy. Sir Bernard Lovell inherited a quantity of Army surplus equipment and set up his research station at Jodrell Bank, near Manchester. From 1950 he devoted a large portion of his time and effort in attempting to persuade the government to take the subject with the seriousness it deserved, and to invest in larger and larger radio dishes. Politicians were beginning to learn about the need to support basic research, but they were slow learners and extremely forgetful. (They still are, except when the subject captures public attention.)

The principle of the interferometer

Michelson and Morley used an optical interferometer in the attempt to detect the slowing down of light as a result of ether drag (described in chapter 4). The principle is to generate two beams of light and have them traverse the same distance by different paths, and discover whether or not they are in phase with each other at the end of their long journey. The fact that they are not in phase is indicated by characteristic interference fringes. It is a remarkably sensitive means of fusing together two images in such a way as to remove any gaps or overlaps.

The same principle can be used with any wave source, including radio waves – in this case not just any radio waves, but galactic radio signals drawn from the skies. (This does not imply that some non-human intelligence is at work in the universe, the 'signals' are natural emanations, with no signaller and no message.) The disturbance is not directed at us, but just happens to fall in the range of wavelengths assigned by international agreement for the transmission of radio and television programmes. (I shall deal with the possible existence of extraterrestrial intelligence in the next chapter.)

The radio interferometer has transformed the way astronomical observations are made, and it will soon increase our knowledge of the universe a hundredfold. The idea is drawn from Jansky and Reber. However, instead of one radio dish, we now use two or more – sometimes even 10 or 20. These are strategically placed in an extended open-air section of a traditional observatory, or maybe just a rented field – anywhere so long as the connections can be plugged into a computer. For a large computer – the larger the better – is nowadays an essential part of the equipment. The dishes rotate with the Earth's motion, each disc centred on, moving across and 'mapping' a particular, relatively narrow strip of sky. The observations are carried by communication lines to the computer. Here they are pieced together to make a seamless record of the whole radio picture of the sky. The point of using an interferometer is that the separate dishes survey an area of the source which includes not only the small areas covered by each individual dish *but all the area in between*. The whole field thus becomes the 'lens' of a vast radio telescope.

When a radio source has been identified, the discs are moved around to obtain images of the radio source from as many different positions on Earth as possible. These are then fused by computer to give the shape of the source, exactly as if we were taking photographs of a complex object and merging them to make a single picture. Two or more radio dishes define the limits of the radio-telescope 'lens', so that a very large area of the 'target' can be 'photographed'.

The Cambridge survey of radio sources

The main problem with the radio interferometer is that, because of the difference between the wavelength of radio waves and light waves (radio waves are about 1,000 times longer), the discs must cover an area 1,000 times the size of the optical telescope lens. The Cambridge interferometer was the next major development in the saga of reception of radio noise. It consisted of two discs which were jockeyed about from day to day into prearranged locations in a field near Cambridge. Thus a cumulative 'audiograph' picture of a huge area of the skies was gradually built up. A modest research grant made it possible to construct a large paraboloidal dish made of wire. This was teamed with a smaller, conventional radio dish. Then, as the research proceeded, a third dish was placed on rails for ease of movement. The two stationary discs were 5,000 feet apart, and the third ran on a track halfway between. Every 23 hours 56 minutes, the Earth (and the dishes) made a complete revolution, surveying an oval ring of sky. The middle radio telescope on rails was set to a new position each day for 64 days. The computer was programmed to sort out these images and make a connected map from them.

As the project developed more and more dishes were added. It was interesting to pass this field week by week and see a crop of dishes growing there like enormous mushrooms. The project ran from 1962 to 1965. The result was the Cambridge Catalogue of Radio Sources. In 1974 the director of the project, Martin Ryle, was awarded a Nobel prize, sharing it with Anthony Hewish, also of Cambridge, who discovered pulsars.

In principle, there are no limits to the number of radio telescopes which go to make up the interferometer or the degree of separation possible between them. The only practical limits are the amount of transmission cable needed, the size of the computers available and the expense of the project. So, at present, the talk in the United States is of Earth-size radio telescopes (that is, interferometers), or even larger installations, which would circle the Earth on satellites. These would be separated by huge distances and would beam their observations to Earth by radio (and thus would not require cables for transmission). Their observations would be synchronised at the parent laboratory on Earth and the records made by each dish would be 'cut and pasted' together on computer. With this kind of installation, the conquest of space, and our knowledge of cosmology, could accelerate exponentially.

Gamma rays and their sources in the universe

Three kinds of radiation are given out spontaneously by the element radium as it slowly disintegrates at measured pace. The three types can be separated by a magnet into (what are called) alpha rays (helium nuclei, positively charged), beta rays (electrons, negative) and gamma rays (a form of electromagnetic radiation similar to X-rays, zero charge). The gamma ray is very penetrating, and was discovered by Becquerel shortly after Rœntgen discovered X-rays. It was named by Ernest Rutherford in 1903. The wavelength is shorter than X-rays, so gamma rays are more dangerous to the experimenter. Humans and animals must be screened by lead walls or have special equipment to protect them from chromosome damage, as with X-rays. Millions of gamma rays fall towards the Earth every second carried by photons. Fortunately, they fail to reach ground level, where they would destroy the genes of sexually reproducing animals, including us. They are stopped by an interaction with particles in the ionosphere and atmosphere.

In 1930 Paul Dirac proposed that, in addition to electrons and protons, there existed other particles identical to them, but opposite in charge. This was an inspired guess based only on a slightly amended, more complex form of Einstein's equation $E = mc^2$, modified to deal with subatomic particles. The solution of one of these equations contained a line which indicated that you had to finish by finding the square root of the square of the mass of the particle. Dirac knew from his schooldays that the square root of any number could be either positive or negative. So he deduced (or jumped to the conclusion) that the mass could be either positive or negative. This was rather a bizarre method of assigning a property to matter – courageous even – as there was no physical evidence, only an argument based on symmetry and, to a mathematician, of intrinsic beauty. However, 20 years later, positrons (that is, electrons with a positive charge), and anti-protons (large nuclear particles with a negative charge) were collected from the atmosphere by balloons.

These 'antiparticles' combine with each other in the same organised way as ordinary matter, but in opposite ways so to speak. They yield not 'matter', but 'antimatter'. This is made up of replicas of ordinary atoms, but the constituent particles are of opposite sign to those making up ordinary matter. When 'anti' particles or antimatter comes into contact with the 'real thing' (the particles of normal matter), there is an instant flash of electrical discharge and the matter–antimatter configurations undergo mutual destruction. Both are annihilated, and are replaced by an equivalent amount of radiant energy. Mass is converted to energy in accordance with Einstein's formula. This is one source of the vast quantities of free energy (such as gamma rays) which invest the universe. A lot of this energy is emitted as light (the product of continuous reactions, such as ongoing nuclear processes, in the Sun).

It needs to be emphasised that antimatter has the same status in reality as matter. The terms simply reflect the order in which the discoveries were made by humans. The situation is rather like the status of women. The word 'woman' is itself pejorative, suggesting that we have the 'real thing' first, that is 'man' (or even 'human') and then, as a kind of afterthought, a subspecies which performs only a single noteworthy function – the 'wombman'. In precisely the same way, we have the primary concept matter and the derivative concept antimatter which, the name seems to imply, can exist only as a sort of imperfect image of the first category. As was suggested earlier in connection with the 'Big Bang' theory (and the same applies to any widely accepted view), the conventions crystallise in vocabulary and soon control our very thought processes. So it is with matter and antimatter. It is not quite beyond belief that 'our' universe is associated with a mirror-image (negative) universe disposed elsewhere as an 'anti-universe' still to be discovered (and maybe even visited) by hardy (and stupid) souls.

Other forms of electromagnetic waves or energy packets are generated, often associated with sky objects. Some of these objects are probably 'dark matter' and hence invisible, others are not. These are continuing processes, and we can see the wave-packets in the form of light quanta which are emitted from bodies such as stars, galaxies and stellar dust. The others take the form of invisible radio waves, X-rays, gamma rays, heat rays and so on. Like the light quanta, energy waves are being generated constantly by cosmic processes of which we know nothing, or next to nothing. It is a vast new section of the universe which was discovered only some 40 years ago. It is still very much a novelty, but is of continuing and deepening interest.

The late Andrei Sakharov, cosmologist and once well known as a Soviet dissident, extended conjecture by positing that the universe originally came into being slightly lopsided, in terms of energy. It was supposed to have happened thus. For every particle (or perhaps body) there was a corresponding antiparticle (or antibody). While the original matter of the universe was falling in temperature by millions, or billions, of degrees various kinds of atoms and particles became possible. They emerged in large quantities from the bizarre hot primal 'soup' estimated to be at millions of degrees. For some reason, positively charged particles slightly outnumbered the opposite kind (there was about one extra in each million). Consequently, antimatter totally vanished, along with more than 99 per cent of original matter. Some isolated, local quantities of antimatter may have persisted in some quarters of the universe, even as anti-galaxies. But effectively, the matter we know was left in virtual possession of the field.

The interest of this speculation is that it might account for the persistence of radio noise (maybe even the slight background hum generated by radio, television and electrical equipment). Thus, many believe, in the absence of a better explanation, some part of this 'interference', as well as the invisible energy waves which surround us, is a surviving relic of the origin of the universe. On the other hand, it may simply be a by-product of the universal matter-and-energy changes still going on – a continuous, boring record of the 'universal love-play' of electrons and protons, as it was once mawkishly described. But inevitably, this kind of speculation starts from highly dubious premises. We recall earlier our cautions about substituting 'sums' and 'thought experiments', or stipulations and faulty logic, for accurate observation and objective data. Sakharov's notions irritate me personally, but I would be the first to agree that this reaction is not typical.

Photons – bearers of light and electric power

The photon is a minute energy packet (a quantum) of electromagnetic radiation. The word 'photon' was introduced by Albert Einstein in 1905 in his paper on the photoelectric effect (see Chapter 5). Max Planck had made the idea of the quantum familiar in his concept of radiant heat energy (1900) in explaining his black-body experiment. (You may recall that this experiment demonstrated that heat is absorbed and emitted in small, discrete quantities and not in a regular flow, as we are inclined to think.) The idea that light was maybe complicated spread when Compton showed (1920) that X-rays were made up of corpuscles. The word 'photon' itself was in common use by 1926. The amount of energy 'transported' in each quantum could be worked out by multiplying Planck's constant (h) by the frequency of the wave. Since h is a very tiny number, the amount of energy in a quantum is also quite tiny. Energy quanta move at the speed of light in a cascade of invisible raindrops. There are millions of them.

Photons carry all forms of radiant energy, not only light but (separately) heat, X-rays, gamma rays, radio waves, and ultra-violet rays. All photons travel at the speed of light. The photon belongs to the quark family (see page 127). It is actually what is called a 'boson', having no rest mass and no electric charge. It is a 'field' particle, that is, it 'carries' the electromagnetic fields.

Our interest centres on the types of energy which leave tracks (sometimes invisible) which can be observed by using the appropriate equipment. We have dealt with the obvious one, radio waves. Next in order of importance are the gamma rays and their sources.

Mapping the gamma rays

George Gamov, the Russian–American cosmologist, predicted the existence of a microwave background emanation supposedly persisting since the Big Bang – the notion that the creation of the universe was marked by a cosmic physico-chemical reaction of enormous proportions. Those who believe in the Big Bang prefer to call it a 'singularity' to indicate, among other things, that the human race has only one chance to make good – there won't be any other creation like the last one (they also exude a certain warmth towards the Genesis account of Creation).

The Americans Penzias and Wilson, using an instrument like a colossal version of the old-fashioned, horn-type hearing aid, discovered this microwave 'fossil' in 1965. The hissing was a persistent noise at the wavelength of 7.35 centimetres. Penzias and Wilson were awarded the Nobel prize in physics for 1978 for what was described fulsomely (and inaccurately) as the most important cosmic discovery of the twentieth century. They themselves seemed unaware of the nature of their finding and the support it provided to organised religion.

Due to the phenomenon known as 'Compton scattering', caused by electrons colliding with photons, the energy level of photons goes up as they receive an extra charge which converts them to gamma rays. (It is an increase in energy because the amount of matter decreases.) These rays are projected in their millions, even billions, from galactic sources (see Figure 23) as a result of this process, and a number of others. The most interesting of these, because of the enormous amount of energy produced,

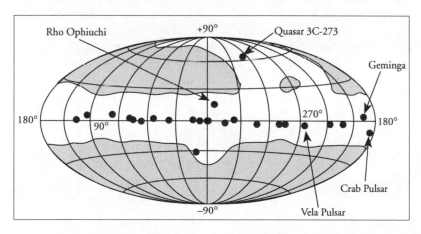

Figure 23   Energy sources in the universe (gamma rays)

is the annihilation of matter as it meets up with antimatter in 'the kiss of death'.

$$\text{electron} + \text{positron} = 2 \text{ gamma rays} \times (\tfrac{1}{2} \text{MeV})$$

(MeV, standing for a million electronvolts – a unit, seemingly of energy, but signifying mass–energy.)

This develops into a chain reaction caused by the mass–energy emitted (equivalent to 500,000 electronvolts per electron–positron union). The chain reaction is due to the ever-increasing number of annihilations, as more and more gamma rays are produced in the repeated reaction shown in the equation. It is another infinite series, which is complete only with the last free proton or electron. The formula, so to speak, 'chases its own tail', in a million-fold repetition of the interaction between more electrons and positrons, producing many millions of gamma rays in a split second. Since 1968 various observation satellites have been collecting information on this, and some gamma rays as well. The most exciting advance being prepared for in this research field is the recording of gravitons, that is, gravity wave–particles, when found. These are similar to light waves; they have the same speed of travel but occupy a different area in the energy chart. When their existence is confirmed, they will form one capstone of Einstein's picture of the universe.

X-rays in the universe

X-rays were discovered and named by Wilhelm Roentgen (1845–1923), a German physician. He found that the bones in his wife's hand could be seen if she placed her hand between the stream of X-rays and the wall. (He tried it first on his own hand – she was cautious – maybe even suspicious. Who wouldn't be?) X-rays have been used for diagnostic purposes ever since, in spite of the unknown risk to our chromosomes. It was discovered that X-rays are being ejected from a large number of stars and galaxies in the Milky Way, in the first instance from the Sun. The rate of ejection to Earth increases as the Sun passes overhead; it falls off again as the Sun moves on. It became clear that it was not the body of the Sun that produced the rays but the corona, which streams out from the Sun and travels for light minutes in all directions. The corona is very much hotter than the body of the Sun. It can be seen best from Earth during a total eclipse, when the solid part of the Sun is blotted out.

In the 1950s a concerted drive was launched to discover the nature and source of 'solar' rays. Unfortunately the means to do so were very limited, rockets being the main research instrument. However a preliminary foray at that time laid the basis for future studies. In 1962

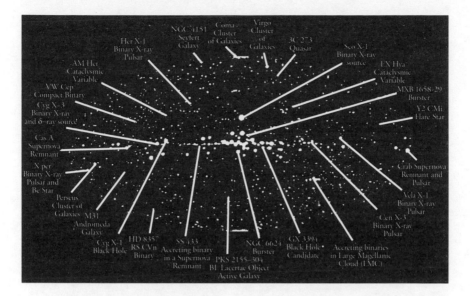

Figure 24   Map of the X-ray sources in the universe

attention was again drawn to the problem by the discovery made by detector rocket that X-rays were coming from outside the solar system. These could be identified by Geiger counters. The sources pointed to were much more active than those normally found. A pulsar source in the Crab nebula was especially remarkable, not only as a source of X-rays but of many other kinds of radiation. The Seyfert galaxies – spiral galaxies whose spectra indicate violent internal turbulence – are also prime sources of rays. Quasars with intensely active nuclei, and some pulsars, stand out particularly.

The next stage in research was the launching of a satellite dedicated to mapping the whole sky by the regular rotation of a detector. The satellite *Uhuru* (Independence) was launched from Kenya on the anniversary of Independence Day in late 1970, as the first of the Explorer satellites. It was designed to rotate on its axis once every 12 minutes. After two years it had scanned the entire sky. Many sources of rays, nearly 200 in fact, were identified. It is now possible to scan these from Earth, so we can dispense with observation satellites. The great clouds of gas between galaxies are additional sources of radiation. The X-ray map conveys the impression that the non-visible (energy) universe is not nearly as extensive as the visible one of stars and galaxies (see Figure 24).

To sum up, the sources of X-rays are active nuclei in galaxies, X-ray binary stars, galaxy clusters, remains of exploded supernovae stars and the Sun's corona. In this list, we have not mentioned the continuing mystery of the dark inert matter that occupies otherwise empty space. In fact, it is possibly a misnomer to call this dark matter 'inert'. It may not produce light, but it may give out a variety of signals. This is perhaps unlikely but not impossible. In any case, even if it is totally inert, the dark matter performs a vital function. It helps to hold the universe together.

Part IV

Human Interest

Chapter 10

The Search for Life in the Universe

Any assertion of the possibility that extraterrestrial intelligence exists is nothing more than earthly hubris.

John McLeish's Thought for Today

Abandoning some cherished illusions

The scenario of the existence of life on other planets found in popular science fiction and in films is pure imagination. Likewise, the stories which appear in the tabloid press about visitors from outer space, and even visits by kidnapped humans to other 'worlds', are fiction. Such stories date from the 1850s, a time when a total baptism in spiritualism seemed to be *de rigueur*. But the underlying ideas have been around for centuries. They resurfaced as soon as the telescope was invented, even before the modern age. Human credulity spread even to scientific circles (especially in physics and psychology) – the twentieth-century gift of the lurid popular press to science.

Confining ourselves to modern times, the literary tradition was started in the seventeenth century by Fontenelle. He was a writer who became permanent secretary of the French Academy of Sciences. He wrote plays, and a famous book, *Conversations on the Plurality of Worlds*, which was published in 1686. Voltaire expressed his disagreement with Fontenelle in his *Micromegas*. The title ('Small big') indicates the major point, namely, his conviction of the minuteness of humans and the infinity of the universe. Can anyone imagine that the universe takes humans as seriously as they take themselves?

We must accept that there is no special dispensation for humans; if there is life on other planets it is more likely to be 'bugs' than people. Bacteria get free transport on meteorites and could readily be dropped off on other planets. Fontenelle, and others since, assume that the human species is so important in the scheme of things that, if life exists anywhere else, then the human species must be prominently represented. This is 'species-chauvinism', a complex of attitudes shared by most humans by virtue of their upbringing. I propose instead to follow Voltaire's insight,

163

with an up-to-date factual summary, placing the theme within a broader context.

Defining life

There are profound differences between living and non-living matter. We do not refer to living 'beings' because this has a specific reference to humans, who have awareness (which is shared with other animals) but also self-awareness, a species characteristic which results from language.

It is often said that, among scientists, there is no accepted definition of life. This is not true. The truth is, and always has been, that scientists act on the basis that certain definitions are ruled out because they do not agree with the facts or because they violate known rules of logic or scientific method. The ways to distinguish living from non-living matter are very clear. Before something qualifies as 'life', every one of its defining properties must be evident, although they may be manifested in a huge variety of ways. In fact, life takes about a million forms. Most of these have been noted in the course of persistent and systematic description by biologists, and the definition which follows is based on this work.

First, an exchange function is vital and must be manifest. The living organism can range in size from microscopic bacteria to hairy mammoths, from germs of scarlet fever or viruses of meningitis to elephants, or to an entire coral reef. Whatever living organisms may be, they always show an intimate relation, an integration, with their environment; it is the prime condition of survival. This is shown by some part of the materials of the environment (edibles) being ingested and transformed by the living organism in a digestive process, with the excretion of waste matter. The accepted, digested matter may be incorporated in the body of the organism as living tissue, or may be converted into the energy which sustains life. The unaccepted is excreted, that is, returned to the environment. There is a special subsystem which processes constituents of the atmosphere. For example, it dissolves them in the water medium which surrounds, makes up and supports all creatures living in this habitat. The digestive and excretory systems oversee all these exchanges.

Second, the process in which excreted materials are exchanged for food leads to growth and change in the development of the organism. Maturation, which includes the development of the reproductive system, occurs. Specialised cells are brought together in the process of sexual reproduction. There is a parallel process of vegetative hybridisation under human control, which acts as a substitute for the sexual method of reproduction and ensures the change and survival of stationary plant species. The many, many mutations of animal and plant life are first 'selected' by the natural process of struggle for the means of existence, or for partners in sexual reproduction. These (spontaneous) changes, if beneficial to survival, are then passed on by reproduction to each new generation. The mechanisms, or essential parts of the process of evolution, were discovered by Darwin (natural selection), and by Michurin (in the Soviet Union) and Burbank (in the United States), who bred across species barriers, breaking the biological 'rules'.

Evolution in Nature controls and directs the superabundance of progeny and mutations. Some of these survive in the struggle for life because they serve to adapt organisms to the vacant spaces available in nature. This system functions only if there are eons of time for it to work, and if the conditions essential for continuance of life exist over the same period. What we know of the solar system rules out the possibility that living cells, as we understand them, are present on planets other than Earth.

Third, there is the property of movement (either locomotion, or body movement in a relatively fixed spot), which is possessed by most (and possibly all) living organisms. They are motivated to search out the necessities of life, as well as partners, to enable them to survive, and to reproduce their kind in a form fit to occupy any niche in the economy of Nature which is available and congenial to their needs. The movement may be minimal, as in plants seeking the sunshine or limpets attached to rock surfaces. (I have never actually seen a limpet move but am sure they must.) Or it may involve journeys of thousands of miles by birds in the air, fish in the sea or human and animal groups on land.

Fourth, living matter constantly exhibits a bias, or 'preference', for the left-hand side, when a choice exists. Complex chemicals such as sugars, which take part in the life exchanges to which we referred, can be made in the laboratory. There is normally nothing to show that the laboratory product is a 50–50 mixture of two types of crystal, one where certain molecules have a right-facing direction, and others which are left-facing. The two are identical chemically, but are mirror images of each other as crystals. They can be separated, one kind from the other. When mixed together, and dissolved in water, the two kinds of crystals have no effect on light. But separately, the different kinds in solution polarise light, causing it to vibrate in a single plane, the first kind to the left, the other to the right. When we test natural sugar from growing plants, it always polarises light to the left. The right-handed mirror-image type is missing.

Living matter often shows this common feature, or bias, where there are left- and right-handed processes. We have seen the preference before where antimatter would seem to have fizzled out, leaving an excess of the 'real' stuff. Where we would normally expect a 50–50 split – because in the absence of life there is a random process – we find a bias towards the left. We live in a left-handed universe, at least as far as life-support chemicals are concerned.

A clear example of the principle of economy of means in achieving results in Nature, a principle most humans have still to learn, is the chemical porphyrin. It is the basis of red haemoglobin in the blood (it carries oxygen and other essentials to vital parts) as well as of the green chlorophyll in plants (which acts in the daily carbon dioxide–oxygen cycle which links plants and lung-breathing animals). It also helps in building up coloured stripes in animals such as tigers, leopards and zebras. This economy of means is shown constantly in the build-up of living nature. It is the factual basis of Aristotle's contention that 'Nature does nothing in vain', supplemented by Newton's statement that, in scientific matters, 'Nature is simple'. It is shown in the fact that, although in theory billions of combinations of organic molecules are possible as 'building-blocks', fewer than 50 combinations of molecules are actually used to build up the 1,500 or so complex organic compounds which occur naturally. Drug manufacturers increasingly add to this number since the number of carbon-based chemicals available for medical use tends towards infinity.

Fourth, 'economy' is a well-chosen word as applied to Nature. Considering life as a whole, there is an incredible proliferation of forms. But these cluster around a small number of common ground-plans. In addition, there are a few large, universal principles which cut across all sorts of barriers. The cycling and recycling of materials is one example. Another is the widespread duplication of a good working design. (For example, the living cell is very similar in ground-plan to the structure of the atom and to the form of the planetary system – a central nucleus with dependent structures surrounding it.)

Lastly, cutting across everything else, there is the second law of thermodynamics. This refers to the system, in this case the universe, as a whole. It states that throughout the many changes that take place, there is an overall loss, or degradation, of energy, and of order (in both living and non-living matter). The law can best be understood if we think of a total fund, or sum, of matter, energy and order in the universe, banked but available. Some part of the system may show a temporary and limited increase in matter, order and energy (living cells are a case in point – they multiply). But the *total* sum of energy is constantly depleted, and can never be replaced, except by an intervention from outside the system. There is no possibility that any process *within the system* could reverse this total movement towards levelling out. In other words, everything in the physical universe conspires, in the long term, to run down. There can be no such thing as 'perpetual motion'. Eventually, the whole system must be without heat, without energy, without motion, without mass. Long before this, it will have ground to a halt. There will be no movement except that of electrons circling the few pieces of matter still around.

The books will be balanced and the account will be closed, not with a final bang, or even with a whimper – just with a little hiss, like trapped air escaping from a radiator. The good news is that nobody will be there to see it or hear it. And it is a very far-off event. Death comes, soon or late, to all parts of the system and, eventually, to the system itself. Nothingness prevails. It is likely that other universes are already forming.

In summary, life is a localised system showing well-marked features. The system operates to produce a temporary and continuous increase in order and an accumulation of energy. But this happens at the expense of the universal environment, that is, everything else. This life is what cosmologists are searching for. The search, led by Carl Sagan, with his many supporters, has become systematic and thorough.

Is there life on Mars?

It is only in the last 25 years or so that we have been able to test a number of hallowed flights of fancy which have gripped the human imagination for centuries. First, we must remove from centre-stage the question of whether humans, or humanoids, are 'out there' somewhere. We need to develop a broader focus on the universe and see it as an environment 'favourable' or 'unfavourable' to life in any form. For example, we need to admit the possibility that life does not necessarily demand a carbon-based (organic) metabolism. Living matter could well be built of sulphur compounds, or with calcium or silicon as the central element. This may be especially true on other planets. Even on Earth, animation can be recognised in vastly different forms of living interaction with the environment. No alternative life-form proposed can be arbitrarily ruled out. The universe is obviously quite different from expectations based on our very limited knowledge. As J.B.S. Haldane once pointed out many years ago, it may be even more curious than we are able to imagine.

Our discussion of Mars begins with the scientific school of thought started by Schiaparelli (1877) and Lowell (1895). These astronomers were convinced, from their telescopic observations, that there was clear evidence of human civilisation on Mars.

It was Schiaparelli, a noted Italian astronomer, and Director of the Brera observatory in Milan, who first raised the matter. In his observations of the planet Mars, he noted certain geometrical patterns made up of straight lines, which, in some cases, intersected at right angles. He reported these lines in the scientific press as 'canali', meaning 'channels', but this was wrongly translated as 'canals'. The Suez Canal, connecting the Mediterranean Sea with the Indian Ocean, had been completed eight years earlier, in 1869. News of Schiaparelli's find captured the headlines of the world press, just at the Canal had done during the ten years it took to build it. It seemed only natural to assume that the '*canali*' were not topographic features, but real purpose-built 'canals'. The mental picture of canal construction was easily transported from Suez to Mars. There was a will to believe, which is not difficult to explain. The story inspired those who already accepted the story of Creation: they took in the *canali* as confirmatory evidence that there were human bodies and intelligent beings ('like us, out there'). The idea was also acceptable to those who couldn't believe that the universe was created primarily as a home for human enjoyment. To many of them it proved that we were not really 'alone' in what they thought of as an alien universe. For others again, it confirmed the idea that God had solicitude for us humans and for people like us. There was something for everybody.

Flammarion, a professional astronomer and the author of bestsellers on popular astronomy (including a book about alternative worlds), accepted Schiaparelli's observations immediately, proclaiming widely that there were, or had been, human settlements on Mars. The evidence, he averred, could be seen by anyone, through even a small telescope. The main importance of his book was to persuade Percy Lowell to change his career. Lowell was a member of one of the two leading families of Boston, the 'Boston Brahmins' as they were called. A 1930s Boston catch-phrase was that 'the Lowells speak only to Cabots, and the Cabots speak only to God'. Percy Lowell, at least, was not like that.

He became an established and distinguished Orientalist, an expert on Korea, Japan and the Far East. Then at the age of 39, he decided to abandon his promising career as adviser to governments, in favour of astronomy. He set up his own well-equipped observatory at Flagstaff, Arizona and proceeded to make some notable contributions. For example, he is usually credited with predicting the existence and position of the unknown planet Pluto, though several others had made similar predictions. The existence of unknown bodies and invisible 'dark matter' in the universe is uncovered by listing unexplained disturbances in the planets' orbits. They are accounted for in terms of gravity acting from some unknown and large mass nearby. Pluto was such a mass. It was discovered in 1930, some years after Lowell's death. Although he had over-estimated its size and weight, it was in the area indicated by his calculations.

In addition to his general astronomical interests, Lowell devoted himself to the study of Mars, about which he wrote several books. His views are of great interest, not merely to astronomers but to ordinary people as well.

Mars is known as the Red Planet. It is an Earth-sized planet in the group of four, Mars, Earth, Mercury and Venus, and it is also the most like Earth in other ways. For instance, it has long been known to have a desert landscape, and the poles are covered in what looks like snow and ice, which melt as the seasons advance. The seasons occur in the same pattern as on Earth, except that each one is about twice as long. This is because Mars takes double the time of Earth to complete its orbit round the Sun. Lowell believed that an intelligent race of beings survived on Mars by using the water melted from the ice-caps by the Sun in summer (he assumed that the snow and ice were frozen water). He thought that the melted water was used to irrigate the desert, being led away from the polar ice-caps by the canals, which according to his theory, had been specially built for this purpose by the Martians. The canals were not just for irrigation; it was believed that vegetation lined their banks, and that this accounted for dark areas along their lengths, as seen from Earth.

Scientific circles were sceptical of Lowell's views in detail, but were prepared to entertain the hypothesis of intelligent life, canals and everything else, until more direct evidence could be obtained. Informed criticism had little to go on, but much was made of two arguments. First of all, there was no evidence of water vapour on Mars even in summer; if ice had been melting there should have been. The second objection was related to the fact that schoolchildren, given a drawing of Mars showing only the main 'channel', and asked to redraw it from memory, more often than not filled in a number of intersecting 'canals' to take 'the bare look off the picture', so to speak. In other words, Lowell's system of canals was labelled an optical illusion. This objection was not very convincing, but it allowed other astronomers to get on with their work and avoid unseemly argument.

There followed a long period of quiescence, with the Martians being mentioned only by the newspapers in the 'silly season' or in Sunday supplements. This kept the subject alive in non-scientific circles and in science fiction. In 1959 the matter was revived by the Soviet astronomer Shkhlovsky. He went so far as to suggest that Mars' two moons, Phobos and Deimos, might be artificial satellites (presumably launched by the Martians). There the matter rested, except among the various scientific 'fan clubs' which flourished around the world.

In recent years, NASA has revived the question, initiating a programme of observation of the planets, intended to last for more than 30 years, using first artificial satellites to photograph Mars, then a crewed spacecraft to visit it. There is also an ongoing programme of monitoring radio signals to try to identify intelligent communications in the Milky Way. These encounters have been carefully devised to settle Lowell's question about advanced human civilization. But there is also the broader question of how to test whether Mars' environment could support life of any kind, even down to microscopic level (but only on a carbon-based metabolism). Preliminary results of this investigation have been published.

More importantly, discussion of the proposed mission has clarified our thoughts about how to look for organic life on such extraterrestrial visits. To some extent, the experiments and observations have made an attempt (not too successful) to correct for earlier geo-chauvinism and the species-chauvinism which contaminates most discussions of these questions. Geo-chauvinism is the human propensity to believe that the conditions we enjoy on Earth – relatively mild temperatures even in polar regions, a life-supporting atmosphere, a beautiful if polluted environment, plenty of suitable food – are 'normal' features of the Earth environment, that this is universally 'comfortable'.

There is a literary tradition that, for some reason, we should be sad to think that we might be 'all alone' in a vast universe. Another one is that there must be intelligent beings 'out there', probably equally distraught by the fact that they have not been able to locate, and communicate, with us. It seems to be a kind of cosmic weariness, akin to the 'Canadian cabin fever', which afflicts lonely trappers isolated and snowbound in the winter. As one well-known science author put it (I leave him unnamed to spare him embarrassment at being associated with such views):

> Somewhere in all space or on a thousand worlds will there be men to share our loneliness . . . Somewhere across space, great instruments may stare vainly at our floating cloud wrack, their owners yearning as we yearn . . . Of men elsewhere and beyond, there will be none forever. [There may even be some women!]

This kind of approach, when indulged in by the Romantic movement, was called the 'pathetic fallacy'. (There is enough bathos there too to call it a 'bathetic fallacy'.) It is a pretence (or maybe even a belief?) that our feelings and responses are shared and mirrored in Nature and by other humans. For example, we are happy and the Sun breaks through the clouds; and everybody smiles to us; or we are sad, and Nature obliges by raining (or crying) in sympathy.

In the case of undetected beings sharing 'our' universe, the fallacy is much more subtle. It is more a matter of mental 'set', that is, a prepared readiness to accept and believe certain happenings to be true, and to twist unrelated events into a pattern which 'fits' expectations. This set is not necessarily bad, for it is part of our human endowment. It is dangerous only when it closes our minds to alternative possibilities based on reality. (The difference between 'sets', of which we are mostly unaware, and moral commitment, which is a deliberate ethical judgement, is small but significant.)

Thus the literary genre of 'creatures from outer space' leads us to think of these 'aliens' as beings with, at least, a semblance of human form. We assume that they (that is, 'proper aliens') live in cities in an industrialised, free-market society, and come from environments much like our own. These ideas are not put into words, but there is a clearly expressed 'set' which leads to them.

The obstacles to true understanding (Francis Bacon labelled them 'Images of the Tribe') become important as sources of error when we plan a close look at other planets, to decide whether they are possible homes for *any* form of living matter. The legal dictum that 'he who asserts must prove', as well as the scientific presumption of an absence of life of any kind until cogent evidence is provided that it exists, must be the basis of our inquiry. (These principles are also essential rules in arguments about the existence of God.)

The Red Planet – close encounters

In the 1960s the whole matter of life on Mars was resolved by a series of satellite launchings and an unmanned spacecraft landing. This was part of the US programme designed to explore the solar system. Inevitably there were a number of technical hitches, but the general conclusions are hard to fault. In 1965 *Mariner 5* sent back 21 pictures of a desert world, showing considerable evidence of earlier volcanic and tectonic activity on Mars. Craters were disposed mainly in straight lines, giving the illusion from a distance that they were canals constructed by humans. In fact, Mars was found to be very similar to the Moon. (By coincidence, this gives an illusion of a human face, an optical illusion fashioned from shadows of the mountains of the Moon.)

Later photographs, from *Mariner 9*, confirm that the terrain is a kind of 'moon-scape' with a number of extinct volcanoes, which happen to be in line because of tectonic, or crustal, weaknesses. (*Mariner 9* orbited Mars for a year, taking thousands of photographs.) There is also evidence of deep, irregular channels which might have been caused by earlier flooding. It is more likely that they, like the rows of the extinct craters, were due to tectonic weaknesses, causing 'Marsquakes.' There is virtually no surface water.

Another interesting finding was that the supposed 'vegetation', which figured in many reports, was also an optical illusion. Mars is covered by a fine dust of red ferric oxide (on Earth, it is known as haematite since it occurs in large blood-coloured nodules, rather like ox-liver. It is not 'rust' as commonly described; this requires water as a constituent, and is a different oxide of iron.) On Mars, this red dust is blown about by periodic, cosmic wind storms which expose the black earth underneath. The fine, red dust is blown back by later wind disturbances. It is this combination of colours – blood-red with exposed areas of black soil – which is interpreted as vegetation. The tectonic earth movements accompanying volcanic activity have produced massive fissures even greater than those of the Grand Canyon in Arizona, while extinct volcanoes on Mars dwarf any mountains we have on Earth.

The data on Mars given below is repeated from an earlier chapter. It constitutes the latest report on Mars.

Mars
Fourth planet from the Sun, uninhabitable

Mean distance from Sun	1.52 AU = 228,000,000 km
Diameter	6,800 km
Mass	0.53 Earths
Density	3.9 g/cm^3
Surface temperature	(-271 to 72) °F, (-166 to 22) °C
Orbital period	686 days (4 seasons)
Atmosphere	nitrogen, no oxygen

Mars appears reddish in our sky, even to the naked eye. In the Martian springtime the surface of the planet seems to change colour as a result of seasonal (cosmic) winds which first cover and later expose the dark surface of the soil. Mars has two small moons – Phobos, the larger, which zips around Mars in $7\frac{1}{2}$ hours, and Deimos, the smaller, which takes 30 hours to orbit. The atmosphere is mostly nitrogen; there is no oxygen. Because of this and the enormously cold temperatures (below blood heat down to unimaginably cold), Mars is incapable of supporting plants or any form of animal life as we know it.

Consequently, experiments to discover whether there is life on Mars were carried out from the spacecraft, since the impossible climate, and absence of oxygen would have been unendurable even for a moment. The results of these experiments are considered next.

Testing Mars for signs of life

From 1965 until the landing on Mars in 1975, NASA (National Aeromatics and Space Administration) prepared for the visit. A spacecraft would settle the question whether or not there was extraterrestrial life. Considered as a first, exploratory visit, the trip was a complete success. Arguably, the definition of 'life' was restrictive, as the crucial experiments covered only forms of metabolism (life) based on carbon compounds. For most people, it is natural to define life in terms of our everyday experience. This is true even of biology textbooks. However, it is a fact that in extremely marginal environments, even on Earth, other forms of life are found. For example, living matter adapted to boiling water conditions is found at vast depths on the ocean floor. The boiling water occurs around the junction of the tectonic plates which float isostatically, as though on water, but actually on the molten magmas of the Earth's rocky and deeper interior regions. Species of spiders, crabs, 'daffodils' and rat-tailed fish have been photographed in these deep waters. Obviously, there is no sunlight at these watery depths, so photosynthesis does not support life. The main food material is bacteria, but there is also a sulphur cycle in the food chain. We can visualise other life systems, and even more exotic forms, suited to the extreme conditions to be expected on the otherwise 'uninhabited' planets. Any life there is probably based on elements other than carbon, hydrogen and oxygen ('organic' elements), which are essential to life as we know it. It is a question of being prepared for the totally unknown and unexpected. The limitation of possible exchanges to carbon-based metabolisms is too human-oriented.

The experiment that was devised to test whether there were any signs of carbon-based life on Mars was very ingenious. It made use of an automated collecting system (a robot shovel) and an automated laboratory. The robot arm shovelled up a sample of Martian soil and delivered it to the chemical laboratory. There, under aseptic conditions, it was inoculated with radioactive carbon 14. This served as a 'tracer' element to determine whether there was, or had been, 'life', even of microscopic size, in the soil sample. The doctored sample was then incubated to 'hatch out' any dormant life which might be present.

In three sub-experiments with this Martian soil, it was concluded that there were no positive indications of any organic material. Doubtless, many more on-the-spot investigations are planned to eliminate any remaining uncertainty. One thing is for sure: there was no trace of the community of canal-builders proposed by Lowell, and there are no canals. Indeed, it has been calculated that if all the moisture on Mars were collected from the atmosphere and condensed as free-flowing water, there might be just enough to fill one of the smaller Scottish lochs.

The reasons for the misconceptions about Mars are also now obvious and indisputable. No organic matter could survive for long because of the fine dust storms. Also the absence of water (the ice-caps are frozen methane) and the extreme, constant low temperatures on Mars could not support life. Moreover there is no protection from the Sun's rays: no trees, no vegetation, no ozone layer. The atmosphere is 90 per cent nitrogen, and there is no oxygen. Certainly, Mars would be totally unable to nurture the beautiful, slender, hyperintelligent humanoids with whom past imaginations have embellished it.

Is there any extraterrestrial life at all?

The vastness of the universe, recognised only in the twentieth century, has diverted attention from Mars as the most likely meeting place for humans and other 'minds'. Once the idea is accepted that life might be discovered somewhere else in the solar system, or maybe in the wider universe, we need to estimate the chances of finding it, if we were to devote massive supplies of money and talent to the search. The probability of finding life in a particular place may be remote, but it has to be multiplied by an almost infinitely large number of cases, which at worst makes the chances fairly reasonable. There are literally billions of stars in the universe, so the number of possible cases of life in other galaxies is very large; in fact it may run into millions.

However, the chances of finding life are not the same throughout the universe. Evolution is a mindless, natural process. As such, it has endured (on Earth) over countless geological ages to produce only a single intelligent species capable of abstract thought and of communicating the results of their cogitations to other beings of the same kind. Now, the Milky Way and planet Earth are relative newcomers in the universe. Many 'places' (in galaxies) have been available to living intelligent beings for billions of years more than we can imagine. There might be many places, scattered throughout the universe, where intelligent life would be welcomed, at least by a friendly environment. This is, of course, a leap of faith, as there is no evidence for it. But ringing all the possible changes over time, it is conceivable that intelligent beings have been delivered in other places besides Earth.

Talking of probabilities, we can carry the argument a stage further. There is a certain low form of life in mathematics known as the Green Bank Formula. It was proposed by F.D. Drake and given the ecologically appealing name of a place in Virginia. I call the formula 'low life' because just about every number needed for the calculation can never be known. Our resources are quite inadequate to gather the information needed, and probably must remain so for all time. The intent of the Green Bank formula is to calculate the number of planets in the universe which could have highly developed cultures right now. Of course, we don't know as a fact that there are earthly planets outside the solar system; nor is there any way of knowing the numbers of hypothetical planets which might have a scientific culture similar to those of the advanced civilisations on Earth, and who might be interested in learning that there are kindred souls in other parts. The problem is that, for a very far-off place to receive our radio message, and for us to obtain a reply, will take many centuries – maybe as much as 100,000 years. This is because the velocity of radio waves is the same as the speed of light. The Green Bank formula is given below, after issuing the caution that it really belongs in the Shakespearean category of 'quips, and cranks and wanton wiles'. In reality, it has the same status as the ploy used by Christopher Columbus to persuade Ferdinand and Isabella to finance his trip to the Indies 500 years ago – which landed him in the New World.

The Green Bank formula

The formula provides us with an estimate of N, that is, the number of potential respondents to radio signals deliberately beamed from some location on Earth, and requesting the favour of a quick reply.

$$N = R_a \times f_p \times n_e \times n_l \times f_i \times f_c \times L$$

where $\times$ stands for multiply, and

R_a = average number of star formations in galaxies
f_p = portion, or fraction of stars with planetary systems
n_e = mean number of planets ecologically suitable for life
n_l = mean number of planets with established life forms
f_i = fraction of such planets where intelligent life exists
f_c = fraction of planets able and willing to communicate
L = mean lifetime of a technical civilisation.

The formula starts from the assumption that each of the stars (numbering billions) in the galaxies could represent an island solar system very similar to our planetary system. There is, of course, no direct evidence of any planets existing, apart from those we can see in the solar system. Using the Green Bank formula for the solar galaxy (the Milky Way), and substituting reasonable guesses for the values of each of the factors mentioned, we discover the sad fact that the answer for our galaxy is – ONE. We are apparently the only people who are distraught and lonely in the whole of the Milky Way. The good news for the formula – or perhaps I mean the bad news – is that, at least for the solar system, it seems to work.

We now apply the same formula to the whole universe. This suggests that an average of one planet per galaxy may have inhabitants of the type specified. We further assume that, hopefully, other (hypothetical) civilisations, being older and perhaps wiser than our Earth-bound culture, would not have devoted 50 per cent of their energies, inventions and technologies to the science of killing each other, as we have done. If, in consequence, these civilisations managed to survive, it might have the effect of vastly increasing the value of L. Thus, we could expect large numbers of potential clients for our good news. The estimates of N (the number of qualified civilisations likely to be involved in the proposed jamboree) is between 100 and 100 million. The range of these estimates is the clearest indication that we have scarcely advanced further than a quick glance through a telescope in the direction of a solution to this problem.

Personally, I don't believe that this formula has achieved anything except to list the steps in understanding which we still have to make to pinpoint the extreme unlikelihood of a positive response. The formula's main use, if it has one, is to keep our attention on the main issues in trying to forecast 'intelligence' (meaning *Homo sapiens*) elsewhere. In any case, there seem to me to be more serious Earth problems than offering psychotherapy to aliens in far-off galaxies.

Radio contact with outer space

In 1960 Drake set up Project Ozma, named after the wife of the Wizard of Oz. (Frank L. Baum was a cult hero for many American intellectuals at the time, as Lewis Carroll had been for a previous generation of English astronomers.) The radio receiving system was pointed towards two stars, epsilon Eridani and tau Ceti, to monitor possible intelligible radio signals. These stars are only about 12 light years from Earth, so it would take radio signals only 12 years to arrive there from Earth, and another 12 to receive a reply. The two stars were chosen as being only about 10,000 billion miles (16,000 billion km) away. The project was secret in the beginning, but became the model for a network of 'listening' stations, set up to monitor extraterrestrial broadcasts from sites throughout the universe.

Before outlining the stages in implementing this plan, we should be aware of the scepticism surrounding prospects for its success. For nearly a century, radio and television signals have been beamed out from hundreds of stations on Earth, and when we switch off our receivers, it has the same effect as if we had withdrawn physically from an international conference. We could no longer follow events, but the conference would go on regardless. Once the energy waves have been activated at source, they travel outwards, and cannot be called back. The transmitting station is the central point of an expanding sphere of energy waves, (the opposite of the CIA's 'cone of silence' which, we are told, descends over conferees to block the sounds of their conversation). It is like a CB radio opening up communication with all and sundry. The waves of energy which carry the programme (which *are* the programme) are an invisible sphere of energy whose centre is the transmitter. We can tap into this source on Earth only because we have the correct equipment tuned in at the wavelength of the broadcast. Provided that the energy of the signal can be boosted sufficiently, there is no site in the whole universe that cannot receive and enjoy the broadcast if the receiver there is adjusted to the correct wavelength. The main problem is the power necessary to boost reception at extremely distant points. The law is, that the energy of transmission (which is transformed into sound and pictures) is spread more and more thinly as the distance apart increases. It does so in accordance with the inverse-square law, falling off as the square of the distance (like Newton's law of gravity and Coulomb's law for electric charges.) The energy 'spreads out' in all directions, so its sphere of action is, in effect, the internal volume of a sphere.

The point of this lengthy explanation is that, if there are extraterrestrials 'tuned in' to receive our signals, they must have been receiving our TV and radio programmes for decades. And if so, it is unthinkable that they are waiting for a signal which the *senders* imagine will be of extreme interest and significance to them. Had they felt the urge to communicate, they would have been clever enough, and have had the technology and facilities necessary to do so. This whole experiment is another variety of the pathetic fallacy, the belief that every intelligent animated body in the universe has a mental and emotional content the same as ours.

When the 'Green Bank' people (that is, the National Radio Astronomy Laboratory) thought about what radio wavelength they should monitor to receive 'alien' broadcast signals, Drake decided on 21 cm, because this is the wavelength of the dark line in the spectrum of solar hydrogen, the commonest element in the universe. This frequency is no worse or better than any other, but its choice over others betrays the anthropocentric attitude we have already described. (Guiseppe Cocconi and Philip Morrison of the Massachusetts Institute of Technology chose the same wavelength but for a different reason: because this part of the radio waveband is relatively free of other users. They tuned into a lot of interference but, like Drake, did not receive any sensible communications. Their project was a great advance in conceptualisation of the problem and its solution, narrowing it down to a specific and appropriate question to which they received a cogent, if silent answer.)

To all this depressing chronicle, Soviet astronomers added a temporary glimmer of hope by noting that a quasar radio source, catalogued as CTA 102, seemed to be sending out intelligible signals. Unfortunately, this could not be independently verified. It was reckoned that this, like other Soviet reports, was probably a mistake, and that the activity recorded was a flare-up of the solar corona, or maybe just sunspots.

The next flurry of interest was caused by a 24-year-old postgraduate student at Cambridge who, in 1967, noticed that the monitor equipment printed out a strange activity in the heavens. Jocelyn Bell was studying radio stars in the Vulpecula (Fox) constellation, when she accidentally made 'the discovery of the century', as it was described at the time (if anybody can be said to have done this, it was Einstein in 1905). The record in Cambridge was showing a regular star pulse every 1.33728 seconds. The Cambridge group could find no explanation for this, and proceeded to eliminate all possible local sources such as electric hedge-clippers, cars passing with their heating systems going full blast, vacuum cleaners, refrigerators and so on. The regularity of the pulse automatically ruled out these possibilities, but there are no sceptics like scientists who have a possible and – they believe – better alternative explanation. The group even adopted one of Jocelyn Bell's facetious suggestions, or pretended to, that it was 'little green men' trying to signal their plight to the Cambridge students. Eventually it was decided that the pulse was coming from a new kind of star, a 'pulsar'. The first one discovered was honoured by the catalogue number LGM 1 (after its imaginary progenitors, and with an ironic side-swipe at Green Bank). About 500 more pulsars have since been identified, and the cause of the pulsations has therefore been explained – needless to say, with no contribution at all from Little Green Men.

In the 1980s a movement calling itself by the acronym SETI (Search for Extra-Terrestrial Intelligence) was established. The astronomer Frank Tipler was probably typical of dissenting opinion, when he pointed out that, since half of the stars in the galaxies are older than 6.3 billion years, we need not bother sending signals there, as 'their spaceships should already have landed'. However, in spite of this note of realism, by 1983 we were able to monitor 130,000 radio channels. A portable 'SETI suitcase' has also been invented. It can be plugged into any existing radio telescope without disturbing normal observatory routines. (There is a new slogan in radio astronomy, 'Have SETI, will travel!')

The whole field is a flourishing branch of research with plenty of money available. Project development has reached the stage of planning to be able, by the end of the century, to reach out to monitor eight million channels, simultaneously covering all radio frequencies. The expenditure is justified on the basis of preparing for invasions from outer space. Economists used to have a name for this kind of thing: 'conspicuous consumption'.

Chapter 11

In Place of a Summary

Young man, in mathematics you don't understand things, you just get used to them.

The great John von Neumann (overheard, to a student)

Mathematics may be defined as the subject in which we never know what we are talking about, nor whether what we are saying is true.

Bertrand Russell

Cosmology and speculation

Since about 1900, there have been exceptional advances in cosmology and related sciences, with remarkable breakthroughs in four areas: the study of massive objects, such as galaxies in motion; the nature and characteristics of the submicroscopic materials of the universe, including fragmentary fission products of the atomic nucleus; the relations between matter and energy; and the study of the cosmos as a system in evolution. The new physics at the base of cosmology began with Max Planck and Albert Einstein. It was then transformed by the theories of Bohr and Heisenberg. The centre of interest then changed to particle physics and the fission of the nucleus, producing an incredible variety of particle types. From the 1950s to the 1970s problems of representation of the main forces of the universe became the issue.

Cosmology can usefully be broken down into four epochs, each corresponding to a particular dominant world-view. In chronological order, these eras are ancient, classic, modern and recent. Ancient includes the pacemakers, Chinese, Arabic and Greek, to the extent that they were caught up in the Western medieval synthesis. Classic refers here not to the Greek and Roman eras, but to the physical and cosmic thinking founded on the assumption that the universe works by causality, whereby events are explained as the outcome of precedent causes. This classic phase includes the 'founding fathers' of cosmology, Galileo, Newton and Einstein. The third phase was initiated by Bohr and Heisenberg, who believed that our observing instruments interfered, to make observation of subatomic events impossible: observation altered reality in a totally unpredictable way. A single observation of the position of a particle, they said, immediately scrambles the values of all the other variables that we wish to measure. This period was dominated by the battle of opinions over the nature of the object observed and the observing subject. The fourth era, which took hold after the World War II is characterised by massive instrumentation and expense. Cosmologists are involved in a race comparable only to the 19th-century race for colonies. In the process, totally new perspectives of understanding have been opened up.

Causality implies realism. We believe that, as a precondition of our attempt to examine their changes and relations to each other, material bodies really exist. This reality is independent of human perception of them, of human wishes and actions. In this sense, psychical and 'psychic' states are not causal, but subjective. Events are due to the action of universal laws of which causality is the chief one. Humans can only intervene to alter this by means of guidance and control devices – a mechanism. In doing so, they introduce another causal factor which affects the outcome. It is a valid extension of these principles to recognise that subatomic events, just like changes of state, are subject to a different pattern of causal laws, but this does not abrogate the functioning of the general principles of realism, including causality.

Aristotle blundered hugely in not recognising the operation of laws of probability. This taunt runs to the heart of his logical analysis, where all things and processes are two-valued: 'either A', 'or B'. Such two-valued logic rules out the category 'C,' having aspects of both. The omission is clear in the problems which arise when we study and seek to explain 'change', 'becoming' and even 'movement'. Parmenides and Zeno, compatriots of Aristotle, posed these problems in the shape of paradoxes. Zeno's paradoxes continued to be mysterious until resolved by the advances of Newton in science and Hegel in philosophy. Since the invention of statistics – and in some cases, since before that – scientists have been thinking in a three-valued logic. They recognise categories such as 'probable' and 'unlikely'. They also use a numerical calculus, along with the category system, to measure degrees of association and probable causation.

Acceptance 'in the absence of proof' is a feature of religious faith; rejection is the mark of scientific belief. Science works in accord with the legal principle that 'the one who asserts must prove'. Classic physics asserts that there is one universe, the one which we perceive. Quantum theory, by contrast, allows us to entertain the possibility that this is not so: there may be other universes beyond our ken, or maybe this universe is a sham, made up of human perceptions, as Berkeley and Bohr contended. Quantum physics is concerned with the non-existent, with potentialities of existence, with virtual particles and vacuums filled with hypothetical matter; it awaits the birth of the alternative possibility.

A serious argument has been made that, if there were no animate life endowed with vision, there would be no such entity as light and colour. This is true only in the sense that vision (of light, or shape or colour) depends on the existence of a functional nervous system. This converts energy, of certain known wavelengths impinging on essential structures and inbuilt connections, into sensations of light, shape and colour. But if we abolished eyes altogether, the various energy wavelengths would still exist. Objects do not disappear if or when there is no one there to see them, nor do they suddenly emerge at the moment we turn our gaze on them.

We cannot glibly transfer the principles we discover in the subatomic state, even when true, to the macrocosm of substantial matter with its special rules of order and relationships. Quantum mechanics deals with processes which involve the interactions between the observer–subject, the observed object and the observing instrument. Previously, it was the physical object alone which was considered as the centre of interest; nothing else was admitted to play an active part in the observation.

Matter, fields and energy: a 10-point summary

(1) It is said that 'fields', not 'matter', are the 'substance' of the universe. But what difference does it make what we call it, since fields and matter are interchangeable? The difference is purely in the image which present-day science likes to project, one word being ideologically better than another. In any case, we should be talking about 'energy–matter', because the category of matter also includes 'substance–matter', made up of particles such as atoms, nuclei and electrons, as well as 'particle-matter', which consists of numerous short-lived, often virtual, transient bundles of energy.

(2) It is paradoxical to speak of subatomic particles in terms of interactions involving mass, velocity and momentum, as though they were automobiles or artillery shells. Yet, some theorists say that particles have no independent existence and are merely manifestations of interacting fields. This is one of several basic difficulties which lie at the very foundation of quantum theory. Paradoxes of this kind – that is, factual paradoxes, not just verbal tricks – are more acceptable in science than elsewhere. They indicate the need for new thinking by embodying something which sounds like nonsense but enshrines the possibility of a new truth. (Warning: Not all nonsense is hereby amnestied.)

(3) At more than 99 per cent of the speed of light, the mass of a body or particle increases dramatically and exponentially. In other words, it multi-

plies, as if out of control. At 100 per cent it reaches infinite mass according to the formula, but this may indicate merely that the extrapolation from normal speeds has become invalid and a new rule has taken over.

'Infinite mass' here probably indicates that a change of state has happened, that is, that different rules take over. It is like crossing a US State line: federal laws and jurisdiction become effective, suspending local enforcement of relevant state laws. In science, the laws of conservation of matter and energy, the second law of thermodynamics, and similar laws of scale, correspond to federal laws in that they can always take over, where conditions are met. Presumably the change in this case means that substantial mass changes into energy – but, at this point, any guess is as good, or as bad, as any other. We can conceal our ignorance with the word 'infinity'. Cosmology has still to come to terms with infinity, even dealt with piecemeal in a number of specific contexts such as the above.

(4) Massless particles have zero *rest mass*, that is, all their mass–energy is the energy of motion. For example, photons cannot be slowed down or speeded up since they have no substantial mass for forces to act on. They simply travel at the speed of light – in fact, they *are* light. So, we cannot draw them in a diagram; indeed they are thought by some to be nothing but a single element in a mathematical structure – whatever that may mean.

(5) If we could actually *see* energy, the 'package' in which it comes is so fantastically small as to be totally invisible. At this level, energy differences are so tiny as to escape notice until they accumulate. On the atomic level however, such packets are the dominant characteristic of Nature.

(6) Most subatomic particles live for much less than a millionth of a second and are too small to be observed directly. Bubble chambers and other detectors are used to identify and count some kinds of particles. Various indicators tell us what they are. One chemico-physical reaction will serve to picture their behaviour and clarify their meaning or concept as 'virtual particles'. The reaction shown is between a π-meson particle and a proton. The latter acts as a kind of catalyst, and is recovered unchanged at each round and at the end of the chain reaction.

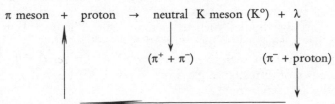

This reaction is a circular chain.

The equation sums up a chain reaction where particles exist for only a minute fraction of time before being caught up in the ongoing process. The reaction proceeds instantaneously. It is a bit like the the schottische, a country dance where partners appear as if from nowhere for a brief second of glory before returning to a kind of musical lockstep which is the continuing round of the dance. Like particles – metaphorically speaking, now you see them (the new partner or particle), now you don't.

(7) Subatomic particles are not made of energy, they *are* energy, better described as 'mass–energy'. They are engaged in an unending, tumultuous dance of creation, an annihilation and transformation that runs unabated within a framework of conservation laws of probability. Within this context, 'particles' are 'tendencies to exist'.

(8) A useful model which helps us visualise energy moving as waves and as quanta (packages) is the two forms of domestic water supply. Water is used in the home in two ways. One is delivered when we turn on the kitchen tap, the other when we flush the toilet. If we turn on the tap and forget to turn it off before the sink overflows, the water meets us as a *wave* on our return. Likewise, the water in the toilet tank is renewed when emptied, and passes for a time, on flushing, through the sewer pipe in a known quantity. Various devices enable us to convert the original volume in the reservoir into a current, or into smaller quantities. But both consist of the same substance, water.

(9) When a rapidly moving body approaches the speed of light, it becomes more massive (the energy of travel is converted into mass); time slows down; and the body becomes shorter in the direction of travel, front to rear. The effect on time can be explained by the following analogy, or thought experiment. To an observer travelling at the speed of light who looks back at a stationary clock, the hands of the clock would appear not to move, as no light from the clock face could catch up with him. Time would seem to 'stand still'. This is merely an illustration of the process: real proof would demand experiment.

(10) It is a feature of the scientific method to give a clear, specific answer to a variety of questions posed in a particular context. Thus subatomic particles have characteristic behaviour modes quite unlike those of massive bodies such as cannon balls. The two laws which describe the behavior of gases (Boyle's law for pressures and volume, and Henry's law for temperatures and volume) are restricted to gases. Graham's law of diffusion in gases is also restricted, but has the possibility of wider application as diffusion is not, in the strict sense, only encountered in gases.

But none of these laws apply to electrons moving in transistors or to musical sounds transported through the air. The rules and possible behaviour patterns are entirely different. A similar principle applies to light as a phenomenon. In the photoelectric effect, when bombardment by light elicits electrons, light behaves as a particle and as if solid; in the double-slit experiment by Thomas Young in 1802, the light makes interference rings typical of a wave motion; in Compton's 'scattering' experiment light behaves both as a particle and as a wave.

Chapter 12

Science and Religion

The last appeal in all researches into religious truth must be to the judgement of the human mind.

Rev. James Martineau

Two things fill the mind with new and ever increasing wonder and awe, the more often and seriously we reflect on them – these are, the starry heavens above and the moral law within.

Immanuel Kant

The religious origins of Western science

The French Catholic theologian Étienne Gilson devoted a large part of his 1932 Gifford Lectures (on the spirit of medieval philosophy) to an exposition of the differences between Being (with a capital B), and being (with a small b). It was fascinating to discover the many shades of meaning this slight spelling change made. Gilson's argument was not just about spelling, but was a serious, concerted attempt to discover the differences between God and other beings. At first sight, to the modern reader, the centuries of argument over such questions by European Doctors of Divinity seem to be of little import to the affairs of the world, and to contribute nothing to the history of thought. The arguments, filled as they sometimes were with theological animus and venom, might easily be dismissed as sectarian quarrels, reflecting only power struggles in the Church, such as have continued to motivate Dominican and Jesuit power-brokers.

However, the central themes of Gilson's lectures were the nature of knowledge and perception and how the antecedents of the rules of scientific method are established. Thus *a priori* principles and logical analysis (reasoning itself) are called to account and subjected to *a posteriori* tests based on scientific objectivity. Logic was testable by experience, if one had the wit to try and the strength of mind to persist. For example, Zeno's argument that an arrow shot into space remains suspended in mid-air because motion was a contradiction could be laughed out of court, on the grounds that experience always has a veto over stupidity, if only one is willing to use it.

Until the success of Darwinism in the 1860s marked the beginning of the end of the long prehistoric period of natural science in Western Europe, it was the general theological context, dominated by Christian fundamentalists, which alone monitored the increase and spread of knowledge. Even in this context, science could not be discussed directly, as a centre of interest in its own right. There was a hidden dimension in the disputes about the nature of God, truth, logic, method, and the process that we now call concept formation.

Like other scholars in medieval times, scientists adopted the establishment concept of a Stairway of Creation. This means that a regular progression could be seen in the natural order. It started at ground level, from the simplest beings, and moved by steps, through levels of complexity, to the Heavens. Moving upwards from the base, we passed through such levels as single-celled beings, to organised plants and animals, then to mammals, then to anthropoids, then to human beings and finally, at the very top, to God Himself. This was the Great Chain of Being. The concept was an organising principle, supplying the particular idea of Nature as a whole. It laid the ground for what is surely the most basic idea in science and cosmology – that of evolution. The great benefit is that, if we can prove that God exists as Creator and care-giver, or even if we just accept His existence, the entire system is validated.

The Ladder of Creation is our first choice. The second is to accept that none of the proofs of God's existence is satisfactory, but that we are ready to accept *on faith alone* that He exists. This is a respectable intellectual position – but to remain respectable, it requires that those who accept religion in this way cease and desist from dogmatism and the coercing of opinion. They also inherit the duty to restrain those who abuse religion by making it a cloak for intolerance and anti-human policies. Belief in God does not entitle the believer to accept excesses committed in His name. On the contrary, the persecution of minorities by virtually all religious sects casts doubt on the value of the religious beliefs which they claim to hold, but don't practise. It also compromises all quietists and neutralists.

The third choice is that religion is clearly false since belief in justice, compassion and mercy are overruled by the vested interests of Church and State. Much could be said about this. The fourth choice, which is mine, is that there is no way for us to prove God's existence, but that, if He does exist, it is very doubtful whether the attributes assigned to Him by believers are correct. The organised Church's view of the deity is misconceived and corrupted at the roots by human self-interest. We know nothing of God, even that He exists. I am tempted to add that

God knows nothing about us (as individuals) and is not as involved in our daily affairs as some evangelists claim.

The Hebrew myth of Creation – a personal view

If only for the sake of argument, let us assume that we can prove that God exists and that He created the universe. What do these statements really mean? We will restate the question. How does our knowledge of the cosmos help us define a valid concept of the nature of God and give an appropriate response to Him, one which is compatible both with His nature and with ours?

To begin with, we can rule out all animistic deities. The idea of God goes back a long way, but early myths portraying the Deity leave out any truly divine qualities as we now understand them. In their place, we find qualities unacceptable even in humans. Tribal gods are irascible, sponsor genocide, are jealous and petty, and often have other weaknesses such as concupiscence. They are not in tune with the 'good' even as this is defined by humans. Anyone who reads the Book of Genesis, instead of just quoting selected bits, will be both fascinated and horrified by the personality, behaviour and talk of Yahweh. He was an ethnic, tribal god, totally lacking in compassion, mercy and discriminative justice. The God whom Christ portrayed in his words, and more importantly in his deeds, was not this bad-tempered and malicious old man (William Blake's 'Nobodaddy'). We may (and do) quarrel with the 'corrected' institutional image of Christ adopted by his religious followers, and the social values for which they stand. But there are, in Christ, qualities of sympathy, understanding and commitment to justice, all of which are totally lacking in the 'Ancient of Days' and his adherents.

My most excruciating religious experience (apart from a visit to Lourdes as a tourist) was to read, at fifteen, the annotated Old Testament of the Bible. It was in part because of the text but, more importantly, it was due to the reaction induced by the notes of some official but anonymous cleric. Even at the age of fifteen, I wondered about the sanity of someone who could harbour the thought, even for a moment, that the Song of Solomon was a poem celebrating the mystical union of Christ and his Church. The incestuous rape of Tamar by her brother Amnon, also described in the text, was given a similar mystical gloss. (I'm glad to say I don't even remember what it was.) What kind of God could inspire such mindless 'commentary' as this?

God in theology

In the powerful analogy of the Chain of Being, the Creator was the vital link. It was He who forged the chain, animating the whole structure to last through time to eternity. Charles Darwin changed all this by showing that the Great Chain was not just a figure of speech, a metaphor adapted to concentrate our thoughts on God and the subordination of nature and humankind to the Divine Plan. In Darwin's theory, the chain was not created in the course of a single week, as Genesis says. It emerged, and continues to evolve, over vast aeons. It was not a Chain of Being but the Ladder of Evolution. There was, so to speak, a continuous creation. New species and varieties come into being in accordance with the law of natural selection, summarised in Herbert Spencer's phrase as the 'survival of the fittest'.

Under the 'theological hypothesis', to use Laplace's term, the idea that God was the prime link in the chain of being, and its only begetter, implied that He was also the Supreme Being, the exemplar from which all other copies were made. In particular, the group most like the Divine archetype, though still very imperfect, was the human species. This is one form of the (fifth) cosmological argument. The chain-of-being idea was thus the source of the so called 'anthropic principle' touted a few years back as a stop-press account of the cosmos. (The word 'anthropic' is a solecism: its users meant 'anthropological' or maybe 'anthropomorphic'.) The anthropic principle specifies that the universe bears all the signs of being compatible with the existence of human beings, and appears even to 'have been made for them'. This isn't really true of anything but certain parts of the planet Earth. It becomes ridiculous if we include even the other planets of the solar system, never mind the remaining universe about which we know so little. The anthropic principle is chatter, a gross exaggeration nurtured in temperate latitudes. In reality, the environment is humanised – and now dehumanised – by humans. It was not created as a cosy nest but is made so, if possible, by human labour. In any case, the anthropic principle is only debatable as a scientific hypothesis; it does not confer a licence on preachers to dispense a new kind of anthropomorphism (that is, that God is like us, and therefore made the Earth as we would like it to be). Let those who claim to believe in an 'anthropic' hypothesis be sent off to the other planets, with their macabre extremes of heat and cold, poisonous atmospheres and absence of plant and animal life to use as food. Even the threat of doing so, should be enough to convince them otherwise.

To return to the main theme, a better example of the congeniality of the cosmos for humans is the fact that it seems to be intelligible. The knowledge that some humans have of Earth and the solar system seems to show evidence of foresight and planning. It is as though a clever mathematician who knew all about calculus, quantum theory, black holes and matrix algebra were somewhere off-stage, pulling the right levers and pressing the right buttons at the correct time.

The idea that God must be a mathematician is a long-held illusion. It was the basis of the religious delusions of Pythagoras and was also sponsored by Plato. In modern times, it was expressed in the eighteenth century, based on the conjecture that even poor humans can read God's intentions in creating matter, dividing it from spirit, and going on to create humans as special – as sons (and, rather grudgingly, as daughters) of God. We take pride in being different from other accumulations of matter. But the notion of God as mathematician is nothing more than human hubris, Satanic arrogance even. It is as though 'the club' were conferring on the Deity an honorary doctorate in mathematics.

It was easy for the Israelites, Arabs and Roman Christians to develop an anthropocentric model of human–divine relations. This was based on the realities of the patriarchal family (in its Judaic, Islamic and Roman forms). It was also easy to have a parochial image of the cosmos based on the visible and tactile habitat familiar to the ordinary person. Christian and Jewish theistic philosophers, following Plato's lead, were hardly interested in an analysis of matter in the scientific sense. The Platonic slanders on Greek science persisted through the Dark Ages, Democritus being the main 'enemy'. It was taken as an article of faith that matter had only a transient existence; the 'soul', on the other hand, was free and eternal. It was defined as being without mass, and not bound by any of the strictures imposed on the body. It was the 'seat' of our humanity, our intelligence, our emotions and our moral values. All these were at some remove or – in theology – absolutely distinct from matter. Thus any desire to explore, or to concern oneself with the nature of matter, was stigmatised as anti-human, anti-religious and anti-God. Persecution of opinion became the hallmark of what was called religion, and the pogrom became the expression of religious feeling by orthodox Christian majorities. This makes it very difficult to take Christianity seriously as a moral institution.

Jewish Scholasticism was less enamoured of Aristotle than were the Christian scholastics. Corresponding to Aquinas as a religious authority in medieval Christendom was Maimonides (also known as Moses Ben Maimon) for the Jewish religion. However, unlike Aquinas, who was a Dominican, Maimonides had not taken any vows of obedience, and his writings were not officially censored before being published. This was true also of Maimonides' Islamic equivalent, Avicenna. But Jewish scholars could be expelled and persecuted by their local religious communities, as later was the case with Spinoza. In some cases Arab scholars were also subject to tests of Islamic 'correctness' (as was Salman Rushdie recently for his novel *The Satanic Verses*). But the Jewish and Arab 'churches' did not have the same courtly, hierarchical structure as the Christian Church in Europe. (The Reformation introduced a whiff of democracy into the Christian Church, but little more than that. However, at least it broke the 'universal' power of the hierarchy.)

Avicenna (1030) and Maimonides (1190) both reasoned that we can say nothing positive about God's attributes since we know nothing about them. It is not even appropriate to attempt a definition, as this would be an assault on God's infinite power, setting some sort of limit on Him based on our inadequate understandings. If God is ineffable, His qualities beggar any true description. Even to recognise Him as Creator has to be handled with caution and expressed with a number of reservations. Otherwise it makes difficulties for believers such as Newton, who asked whether God could create other universes to fill up vacancies and imperfections in this one.

All we can attempt is what is called 'negative theology', that is, to itemise what God is not. This was pointedly stated by Maimonides, less so by Avicenna. (Aquinas also accepted the idea of a 'negative' theology, but said that it should be supplemented by a 'positive' theology drawn from the Bible and the teachings of the Catholic Church.) We can exemplify the method of negative theology by discussing God's attributes. For example, we might start by denying all human-oriented (anthropomorphic) statements about the Deity and His presumed attitudes. God cannot take any pleasure from being like us, or we like Him. Similarly, He can derive no satisfaction from our constant declarations of submissiveness, or our unremitting requests for aid. He manifests none of our human qualities. (Maimonides cheats a little by ascribing the quality of 'intellect' as one which we share with God. This violates his own principle that God is ineffable, and is especially disadvantaged by any comparison with humans.)

What we *can* do, according to the Jewish thinker, is examine and study God's handiwork and governance (that is, His control, dominion over, and general oversight) of all natural processes. This is what I have tried to do in this chapter, in seeking to apply Maimonides' principles. It is implicit in Einstein's reference to the God of Spinoza, since Spinoza put forward essentially the same views as Maimonides. Maimonides also suggested that special secret knowledge about Nature was part of the revelation given by God to Moses, along with the tables of the Law. (The verses describing these tables (Exodus 33:17–23) are, incidentally, the quintessence of anthropomorphism, but Maimonides does not remark on this.)

Spinoza taught (1662) that God is not concerned with, or involved in, human affairs; He doesn't really 'care' for humans, either as individuals or collectively. Few religious thinkers after Spinoza have followed his lead. However, in the present century, it seems apposite as a clear and logical explanation of Divine inaction in the face of 'man's inhumanity to man'.

(That there may be some kind of 'divine plan' which accounts for the hecatombs of victims, in the two world wars, in Stalinist Russia, in Nagasaki and Hiroshima, and in a dozen other areas, only makes it all seem so much worse.)

There was an intellectual and popular reaction to the Christian theologians' declaration of their monopoly on opinion and belief, but this dissent was suppressed. In defence of their special status, the theologians produced various 'arguments' to counter unbelief. These were developed into a sustained, logical discourse by the scholastics. Beginning with Saint Paul, this flourished from the early Middle Ages and was crowned by the work of Thomas Aquinas in the thirteenth century. However, our scientific understanding of natural processes has cast doubt on the general spirit and relevance of theology today. We are now more concerned with human problems of survival, the economy, and social issues of discrimination, than with intellectual arguments about God. To use modern parlance, real public discussion of religion has been put on the back burner. Whether, however, the 'proofs', or arguments, in favour of God's existence are convincing is a matter for individual judgement.

Arguments for God's existence: a critique

The central questions are, does God exist, and can we prove it? By tradition, there are four or five 'proofs', which attempt to answer both questions positively.

The first proof is known as the ontological argument (that is, the argument 'from being'). It goes thus. We all have a notion of God. This notion is of One greater than anything or anyone else. But what exists in fact is greater than what exists only in thought. Therefore God exists.

This argument was not well received even when first put forward (by Saint Anselm of Canterbury, d. 1109). It was taken apart word by word by learned disputants over the centuries. It has long been recognised as a mere logical trick, or spoof. Its chief value is that it alerts us to be on our guard against contemporary cosmologists who engage in rather similar semantic or, more often, mathematical games.

The other arguments have rather more intellectual content, and are more plausible. The second proof was derived by Aquinas from Aristotle's treatise on physics. It is the classic proof from motion. We can observe that things are moving, it says; in fact, motion is a leading feature of the cosmos. It is also clear that this motion is transmitted from one body to another, for example, by collisions. But, to Aquinas and Aristotle, it was equally obvious that there must have been a first mover, else there could be no second or third mover, and so on. An infinite regress is inconceivable in philosophy, and hence impossible. (The Greeks were ignorant of infinite series and, of course, radioactivity.) So we cannot go back in time unless there is a first mover. This Prime Mover (now endowed with capitals) is what we call God.

It should be noted that Aristotle, being a pagan, did not emphasise that it was a personal god who created the world, though his writings strongly suggest that he was a monotheist. But a great number of forgeries in his name circulated in the Middle Ages, and his writings on the nature of God may have been severely edited before Aquinas ever made use of them (it was known as 'pious' fraud).

To prove his case, Aquinas adapted Aristotle's notions of the First Mover and the impossibility of 'an infinite regress'. He already believed with absolute certainty in a Creator. Now he tried to correct some off Aristotle's inconsistencies and errors. In fact, Aristotle was wrong in his idea of the need for a 'first mover', and he was wrong in believing that an infinite regress was impossible. Aquinas extended the concept of infinite regress to refer not merely to numbers (which we now accept as infinite in several ways), but to dependent categories in any sort of classification. For example, unless we have a fundamental category (such as 'matter') there can, in logic, be no sub-categories (such as 'compounds', 'atoms', 'protons' or 'quarks'). Everything would be in a primitive chaos. This argument remains open to criticism as being still chair-bound in the deductive Aristotelian mode.

The third proof is constructed logically in the same way. It follows from previous arguments that God must have been the Creator of all that exists. If not, then there must be 'another', who will *ipso facto* be superior to the first god. This argument is logically sound, but it says nothing about the existence of God, merely that *if* He exists, then He is supreme.

Fourthly, we have the teleological argument. This is Archdeaeon William Paley's famous 'argument from design' (1802). It figured prominently in the evolution debate in the 1860s, becoming a central issue in the acrimonious quarrel about evolution between scientists and religionists. Paley's reasoning is that we would normally explain the intricacies we find on opening a watch-case as being a contrivance devised by some craftsman. It shows evidence of having been meticulously planned to perform the special function of marking the passage of time. And in exactly the same way, the diverse structures we find in nature, such as those which ensure sexual reproduction in orchids, or the mechanisms which elicit instinctive behaviours in animals, point to the existence of a Designer or Master Craftsman (with supreme power over nature). This Designer is what, or whom, we call God.

This argument has always been the most compelling in commanding respect. But it was shattered by Charles Darwin in *The Origin of Species* (1859). There he gave numerous analyses of design in nature, as well as showing the seemingly purposeful adaptation of body structures and organisation to function. He explained each of his examples in detail, because he anticipated them being used as arguments against his theory. He managed to prove that, in reality, each was an excellent illustration of the adaptation of organisms over many generations to their environments. This is the most convincing part of Darwin's book.

The protracted procedures which ensured the survival of individuals were part of the very long drawn-out process of natural selection which operates all the time, on each generation and every individual. Thus, creatures are not constructed from a design drawn by an architect, as Paley seemed to imagine, nor from a device which maps them out in advance. The explanation is in the struggle for survival, where the direction of adaptation and the final purpose are controlled, and monitored, moment by moment, by a natural but mindless process. The construction of a beehive is an example of this natural process – no mental plan but a final product which markedly shows clear signs of intellect and purpose, but in the undeniable, total absence of mentation.

How does natural selection work? It is caused by an excess of births over subsistence in nature, in each generation. This excess of birth over available means of existence leads to an unremitting struggle which pits each member of the species against every other. The struggle goes on unawares and without malice or deliberation – it is merely an assertion of one's right to life untrammelled by any consideration of the rights of others. It affects the species as a whole.

Darwin also proposed that there was a natural tendency for individuals to vary from the norm of the species as represented by their parents. Such variations could be an advantage, or a handicap, to the individual in the struggle for survival or for a mate. Thus, beneficial changes are selected, while others die out. The species survives, but changes slightly from generation to generation. It 'evolves'.

There is a complementary natural law of mutual aid which acts within, and across, species. It was recognised by Prince Kropotkin (a famous geographer and revolutionary) in 1902. It is rarely mentioned, as Kropotkin's views and illustrations are disregarded by the dominant free-market thinkers, including most biologists. The law of heredity, that 'like begets like', ensures the passing on of new mutations, and the future survival of the species. Heredity, since Darwin's day, has been developed on the basis of the laws of probability (which were denied as impossible by Aristotle). The science of heredity explains many of the awful physical and mental handicaps which many people inherit and have to endure. It is a feature of human societies that we protect our handicapped young from extinction; in the wild, they more often perish. Neither Darwin's nor Kropotkin's laws operate in human societies unless our legislators design them to do so. In a human society – or perhaps I should say 'primitive' societies – the law of mutual aid acts to cancel out or moderate the law of the 'survival of the fittest'. The teachings of Christ, and of other social revolutionaries such as William Beveridge in Britain and F.D. Roosevelt in the United States, not to mention socialist thinkers, are meant to transform mutual aid from being a law of the jungle into being a social law as well.

We are excused the task of discussing the 'problem' of evil since it is no longer a problem. It has nothing to do with God, but simply reflects, in manifold ways, the failures of humankind. We need not follow Omar Khayyam in his solution to the problem, that it is God who needs to be forgiven, not humanity. Neither are we required to thank Him, or at least to expect Him to acknowledge our thanks. We have to agree with the woman who reportedly said to Thomas Carlyle after one of his lectures, 'Of course, I accept the universe'. We must do the same – and this includes our moral autonomy and social obligations. (Incidentally, Carlyle is supposed to have replied: 'By God, Madam, you'd better!' We take the invocation as an expletive, not as symbolic.)

Countering Paley's (1802) argument from design, Darwin gave a detailed history of the probable evolution of the human eye. The notion that such a precise, fine structure was not planned from the beginning may come as a shock initially. (Darwin's criticisms of its failures in design may also shock. Many people too, may be unaware that precise mechanisms, like Paley's clock or grandfather's gold watch, also have an evolutionary history based on a process of selection for human convenience – one that is an exact replica of natural selection.)

Paley's argument of purposeful design by the deity was rejected by scientists, who normally prefer an explanation from natural causes. This is part of their professional mandate, and has nothing to do with religion. As a result, the existence of compelling alternative views results in a Scottish verdict on Paley's theory, 'not proven'. Scientists do not necessarily agree that an appearence of design points to the need for a designer.

However, the question of design has now been reopened by our new twentieth-century insights. As we pass from an Earth-based concept of Nature to the vast cosmic scale, and while we await an extended view of cosmic processes and a sound theory of origins, the question of design must remain in limbo. The real failure of theism is that it does not give even a whisper of a reasonable guess, either of God's intention in creating the universe, its purpose, or of His actual plan for the human species – the purpose allegedly thwarted by our first parents. Arguments about existence and purpose must remain unconvincing until these questions are clearly answered. What we need is an up-to-date natural theology which takes account of twentieth-century advances in cosmology (and nineteenth-century advances in biology). We need a natural theology not just of words, but one which rejects involvement with inspired texts and prophecy along with the conventional mental sets of the believer. We need a new Cartesian method which starts by doubting everything, and especially 'holy' this and 'holy' that – including, especially, 'holy' ignorance which, following the example of Tertullian, takes credit for its blind faith. The final argument is the cosmological one, put forward several times in different contexts in this book.

Having ruled out revelation as irrelevant until it establishes its credentials, I have to say that cosmologists really know next to nothing about how the universe of stars, galaxies and dark matter does actually function *as a system*. They know the bits and pieces in tremendous detail, but the origins, evolution and function are another matter. 'Initial singularity' is used as a blanket solution to many problems. Polite words about the deity in this matter are normally no more than a cop-out, or a boost for the human species. The statistics concerning what seems to have happened in the first mini-fractions of a second of Creation are all right in providing an intellectual pyrotechnic display. But they advance our real understanding of the universe not one jot.

There is little doubt that the twenty-first century will explain both how the vast material immensity of the universe came into being, and the processes which serve to maintain its existence. It may also discover that we have been asking the wrong questions. As Kant has implied, we don't really need a Creation to demonstrate the existence of God. The strength of the case, according to his way of thinking, is in the moral order and the moral imperative. If God exists, we can be sure too, beyond doubt, that, unlike earthly monarchs we have known, He is not interested in royal protocol, and is not eager to reward us for repeated acknowledgements of His power. Go to the ant, as the Book of Proverbs admonishes us. Can one imagine that God might be 'angered' if an individual ant forgot to utter its gratitude to the Creator for its full daily employment, and for the comforts of living in a united, efficient and healthy community? It remains a moot point whether ants are of less value than humans in the grand scheme of things (if there is one).

It must be said that there are a number of ways of knowing reality besides science. Science is undoubtedly the most powerful means of knowing and our first choice. But, if we use the kinds of safeguards, checks and balances applied in the scientific method, then art, philosophy, poetry, and even political activity and mysticism, have their own essential methods and points of view which round out our understandings. They provide a number of human directions and perspectives, especially in relation to values, with which science has no mandate to deal. However, there is nothing to prevent a scientist, as a person, and a human being, making a stand where his interests and values are involved.

Myth is a way of dealing with the unknown. It is however a treacher-
ous way of dealing with the unknowable, raising more problems than it is
capable of solving. Myth is a concession to our need for *any sort of
answer* to our questions. In important matters of belief and moral values,
we need to be sure of the real sources of our perceptions. In general, a
critical and creative intellect, working on perceived data, is more useful in
the discovery of truth than the possession of an abstract, Simon Pure
scientific method, an impeccable natural theology, or a claim that this or
that brand of orthodoxy is Divine Truth.

Can God exist in a scientific age?

Cosmology has nothing to say about the existence of God – it has no
standing in this matter. What it can tell us concerns assertions made about
His qualities, to check whether these are compatible with what we now
know, and how such qualities are perceived in the modern, scientific age.
Without being negative or legalistic, we have to invite the believer to say
exactly what it is we are being asked to accept. We can narrow down the
area of legitimate debate by accepting, without discussion, some of the
traditional attributes specified as Divine, for example, that God must be
incorporeal. (This simply means that, in His normal form, He is not
limited by having to function, as we do, restricted and bounded by
operating within a material body.)

We can accept that if God exists, He must also, in the strictest sense, be
infinite – without limit of any kind. This is congruent with the first
attribute of His spirituality. This means we need to correct our concept
of the universe and to extend it infinitely in space and (on one interpretation) in time as well. If the universe was created, it must have had a
beginning, though it need not necessarily have an end, whether accompanied by a bang (unlikely) or the final hiss of a tiny radiant discharge
(more probable). It might be acceptable as a mental crutch, although
highly questionable in strict mathematics, to speak of God as infinity,
and the universe as 'infinity minus one'. But however we picture it, the
essence of the matter is that if God is infinite, then His Creation cannot in
any sense be infinite as well. It must be limited in space and time.

It is a feature of humans to change. Also, as humans, we prefer to look
on the happy side of things, so we generally forget to say that change
includes not only positive development and growth, but also decline,
decay and degeneration. Change is thus a single process with two aspects.
We cannot correctly picture God as an angry, fearful Old Man along the
lines of the Old Testament – the 'Ancient of Days'. Obviously time
cannot possibly affect God, who is incorporeal anyway.

Einstein, when speaking of the divine intentions, habitually expressed
himself in conversation in such terms as, 'the Old One does not . . . the

Old One is not . . .' This wry form of words was meant to do several things: to refer directly to Jehovah as perceived traditionally, and especially in Old Testament times (see the Book of Daniel 7:9); to show that Einstein believed in the concept of God; and to indicate a wish to dissociate himself as inoffensively as possible from the crude anthropomorphism and imagery of the biblical text, and with those who took it to be divine truth rather than a series of exquisite, legalistic, annalistic and poetic human constructs.

Chapter 13

The New Cosmology: Future Prospects

Cosmology is often wrong, but it is never in doubt.

Lev Davidovich Landau

Even a state as simple as the vacuum has properties we do not yet understand.

Alan Guth

String models – building a universe

'Euclid defines a point as that which has no dimensions.' In the bad old days, this was usually the first remark made by teachers in class in the first geometry lesson. Then, in their very next remark, they would totally confound the understanding of those who were thinking as well as listening, by adding: 'And I will now draw one on the blackboard.' This was long before the words 'new maths' had ever been heard. In fact what is called 'new maths' was actually invented by the ancient Egyptians who knew about active learning in mathematics long before Euclid and his followers destroyed it.

Instead of a point, which marks position but has no dimensions (which means it is invisible), cosmologists towards the end of this century seriously toyed with the thought that a certain kind of material reality could be a one-dimensional 'string' – a new and rather unusual form of matter. This view has been described as 'the theory about everything'. The geometrical point is the nearest we can get to a concept of the atom, so a line is the nearest we can get to define a string.

The idea of a string is not just a wild guess. It has a certain basis in research reality, but starts from the assumption that the universe started as a super-heated, almost infinite mass of uniformly plasmatic 'soup' which quickly solidified as it cooled. (Gamow in the 1930s called it 'ylem'.) As it cooled, it passed through new phases, it reached new 'lows', and generated new elements, new compounds and new solid objects as it reached lower and lower depths. Things, forces, states of matter, laws and properties, came into being which had never been known or conceived of before. Nobody has yet seen a string, because, if they exist, they appeared almost immediately following the hatching of the cosmic egg. They date from the very earliest micro-seconds of creation. Strings were preceded by cracks which appeared as flaws, cracks due to non-uniform cooling of the primeval soup. This eventually solidified on cooling. The large cracks were filled with the molten residue which hardened on cooling, out of phase with the rest of the material, like veins in rocks.

The first point to be made about strings is that, to make the idea concrete and digestible, the word 'string' was used illicitly. In our reality, in contrast to its Euclidean (geometric) image, twine or string is a three-dimensional material with a certain weight, well within the range of portability. As distinct from cosmic strings, geometrical strings are uni-dimensional. Indeed, they are abstractions defined in geometry like points, except they have one dimension – they are extended, as lines. Cosmic strings, as defined (because they have never been observed, except maybe indirectly), are not too far removed from this abstract definition. They are quite invisible.

Expressed as a fraction, strings are less than a millionth of a millionth of a millionth of a millionth of a millionth of an inch thick (that is 10^{-30} inches). But in direct and extreme contrast to this infinitely small cross-section, they are about 10^{22}, or ten thousand million million million times denser than water. Like points, they may be drawn as diagrams, on blackboards or pieces of paper, but these are merely visual aids. Strings, in the strict geometric sense, have no body, in fact, at least at this stage, they seem to be abstract concepts. Even so, we could be surprised, not to say dumbfounded, if cosmic strings were discovered tomorrow – as we can remember being flabbergasted when 'negative matter' (see below) was found to be an actual physical reality.

Scientists are now involved in trying to account for the four basic forces of the universe, and with the nature of matter into the bargain. This is all quite speculative, and maybe even interesting. So far, I have spoken only about abstractions. We move slightly closer to reality in accepting that cosmic strings might exist, and that the cosmos may have been structured by strings. About an inch of string must have weighed about the same as the Trossachs mountain range in Scotland or the Rockies in Canada. Some of the original strings, from the period when the original singularity had been encountered (if there was one) and the new universe had cooled to the temperature which made strings possible, may still exist as fossils, or relics of the big bang era. Strings may have provided the groundplan, a kind of design or matrix, within which galaxies were laid down in a kind of cat's-cradle string figure.

Strings can be short or long, looped or open. They could account for the extreme amount of 'invisible dark matter' in comparison with the 10 per cent of bright matter visible at night. They pinpoint the need to focus our attention on the dark and invisible substrate of the universe, incorporating it in a completely new picture of the origins and functioning of the universe.

How the formless developed shape and structure

The four forces are the strong and weak nuclear force, the very strong electromagnetic force, and the weak force of gravity. In considering a possible single, integrated theory of force, we have to grapple with still another idea – that is, symmetry. This helps us to work out how things may have come to pass. It gives us a new vision.

The circle, as a geometrical shape, shows rotational symmetry. A circle does not change as we move it around. When we see it in a mirror, the circle is still a circle. It has reflection symmetry as well as rotational symmetry.

The principle can be made even more general. When we recognise all the different kinds of symmetry shown by objects or classes of objects, or theories or processes, and then proceed to change the symmetry systematically by breaking off, or combining particular properties, or assigning them differently – in short, reclassifying our material – we make constructive use of traditional logic. Indeed we are engaged in legitimate thought experiments, a valid use of Aristotelian logic for planning experiments in science.

It is possible to illustrate how this method works, assuming that there really was a big bang. For example, water and ice are very different in their physical properties, they both differ from steam, the third member of the trinity. But these are not new substances, they came into existence in the early stages of Creation when things were really cool. Nowadays, they exist as changes of state, which are linked to temperature. We can heat ice, or freeze water or condense steam, thus changing the *state* of 'water', but not its substance. Either way, what we are doing is to break or abolish one kind of symmetry (liquid water has 'rotational' symmetry, solid ice has 'crystalline', or 'dimensional' symmetry, and steam has 'expansional' symmetry so that it can fill up empty volumes). But they are each a form of water (H_2O).

The theory of strings assumes that, in the very earliest stage of the universe, the temperature was unimaginably high (billions of degrees on any scale). The universe was a formless 'soup', uniform and without structure. There were no laws, either natural, human or Divine; no distinct substances, no Heaven and no Earth. But the universe was finite and this is what saved it. Beyond the finite chaos there was the Void – absolute nothingness, its temperature Absolute Zero. Nothing moved, there weren't even atoms. The main effect was that, in a flash, things began to happen, time began; things, categories of things, forces, and laws took shape and took over.

This, at least is an earlier version of Creation. It is more 'economical' in Occam's sense, using few hypothetical constructs. It therefore seems more in tune with science than later, more mystical accounts.

Gravity lenses

Light can be altered in direction by a shaped piece of glass or by the lens in our eyes, bringing the light to a focus (see Figure 25). From being a beam of parallel rays, light is brought to a point or focus, and then separates out again as many rays. The telescope works on this principle. Newton predicted that light would be changed in direction ('bent') by attraction, as it passes close to a massive body. Einstein confirmed this, but predicted that the deflection would be double Newton's calculated value. This was verified by Eddington in the solar eclipse of 1919, as a 'proof' of relativity.

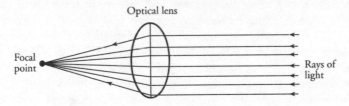

Figure 25   Converging optical lens

The next step was also taken by Einstein who recognised that if we are in the course of observing a far-off star, and some massive body was so placed as to influence the light rays coming from the star, this would affect our perception of it. In certain circumstances, having to do with focal length, the two rays would not coalesce. We would then see not a single star, but be looking at two beams of light. This is due to our position relative to the focal length of the lens. It would give us the illusion that there were two stars. There would be a double image instead of a single image. It would be very unusual, but the heavy mass would in these special conditions act exactly like a gravity lens. As light is transported by photons, so gravity waves are likely to be transported by gravitons which behave very like photons, except that they are gravity waves instead of light rays. The functioning of a gravity lens is shown overleaf.

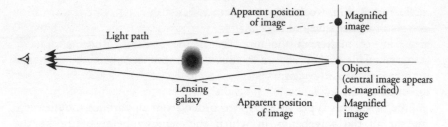

Figure 26 A gravity 'lens': light bending

Light travelling from a distant source to Earth, passing through and alongside galactic masses, will be grossly affected as if passing through a markedly defective convex lens. If the lensing galaxy is smooth and circular, multiple images of the source can occur. If the source is directly behind the galaxy the images can smear out to form a ring known as the 'Einstein ring'. In general though, the lensing galaxy will be lumpy and so distorted images will arise.

Evidences of invisible matter

The above model was proposed by Alan Guth in 1981, and was drastically revised in terms of size the following year by Linde and by Albrecht and Steinhardt. It must be repeated that the numbers, relating to absolutely minute divisions of the first two or three seconds of creation, are chosen to serve the convenience of a particular set of mathematical equations and have no possible representation in reality. They exist (to quote Hawking) 'in the mind of God'. There is absolutely no physical evidence for these minute time intervals, nor can there be. The revised 'inflation theory' corrects some gross mathematical errors in the original (Guth) model. So we must accept, on faith, before we start, that the universe was a special creation. The evidence for this is said to be the background cosmic radiation which, according to Penzias and Wilson, comes on a certain restricted wavelength from all directions in the universe. This is supposed to be a relic of Creation.

During the several millennia in which humans have been observing the heavens, the objects in the skies have been identified as large or small spots of light against the background of a dark sky. Other objects are sometimes visible as moving sources of light – comets, showers of meteors or vast areas of continuing galactic illumination. These are sometimes cut across by dark areas, of cosmic dust or black nebular structures, which may look like bars. On occasion, when the light is just right during the daytime, we may catch sight of the Moon or one of the planets close up, an alien 'intruder' in our daylight setting. But our recurrent celestial perceptions are of the night skies. We normally identify the presence of astral matter by the fact of its shining brightness. Vera Rubin, working at the Carnegie Institute, reported in the 1980s the result of her study of dark matter. We now accept that up to about 90 per cent of matter in the universe maybe consists of dark invisible material. This rather shakes credibility in the accepted cosmological theories of the early universe.

Much of the evidence for invisible dark matter is that it explains the source of various types of energy recorded from all parts of the universe and at all the wavebands in which such energy belongs (unlike the microwave 'fossil' energy from the 'creation singularity'). As we have said, the dark mass is a kind of non-visible (because we can only see matter when it is *self-luminous* when the sun is shining) infrastructure or framework for the bright galaxies and super-clusters of stars which would fly apart and otherwise disintegrate as systems were it not for the enormous gravitational forces generated by the dark stuff.

The most interesting development is the search for gravity lenses to use as a tool for recognising the presence of the incredible mass supposedly distributed across the universe in cosmic strings. Dual images, detected by systematic search for them, where they appear as if aligned along an invisible 'string', compete with cosmic microwaves for our belief. The 'creationists' seem to be in the lead at present, but the matter will eventually be decided by a new generation of cosmologists. Since Penzias and Wilson, and since the 'steady state' theory was temporarily abandoned by Fred Hoyle, opposition to the Big Bang theory has been in almost total disarray. The trend seems to be reversing with the discoveries noted about the actual distribution of matter in the Universe. But there is a long haul ahead.

We can immediately recognise the special case of invisible matter which is secreted in 'black holes'. It exists there as extremely dense, collapsed structures (rather like old crushed cars in the wrecker's yard). It is the solid residue vacuumed off by the intense gravitational pull of the omnivorous, secret, black, disposal plant. The matter here is believed to be massively heavy, but this is all mere speculation at the present time.

Imaginary numbers

A striking advance by scientists in the computer age has been the development of many ideas and methods for dealing with quantities and processes which previously had status only in the realms of the imagination. These methods were not a matter of self-indulgence; they were designed to achieve solutions to real engineering and other practical problems.

The process started in the ordinary classroom with chidren of 14 or 15, the subject matter being powers and roots. Square roots are easy. The complication is that a negative number, multiplied by itself, yields a positive result. So the square root of a positive number breaks down into two equal answers, one positive, one negative. (The square root of four, for example, is plus or minus two.) What about the square root of a negative number? We were once told quite simply that they just did not exist, they were purely imaginary. Ah, yes, you might say, but what do they look like, if only in the imagination?

It was the Swiss Argand who earlier gave us the picture – the diagram named after him. This enables us to work out complex numbers, these being a mixture of a real number (like 20, say) and an imaginary number (like the square root of –3, say). The real part can be plotted as in a graph, as a 'plus or minus' number, on the horizontal axis. The imaginary part (which may also be positive or negative) is counted off on the vertical axis. So each imaginary number is represented by a single point. Thus, imaginary numbers are received into the fold.

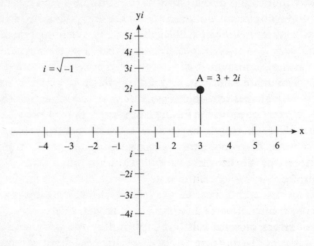

Figure 27   Complex numbers plotted as real/imaginaries

Fractional dimensions of space–time

In the 1930s and 1940s there was a singer in the USA. Her singing was atrocious – and the effect was compounded by the fact that she herself didn't seem to notice this. Her singing became so notorious that she became a cult. People flocked to enjoy the spectacle and maintained the conspiracy that they didn't notice. At one point, perhaps thinking her talent needed re-furbishing, she decided to take a refresher course with a famous trainer of sopranos. Having tried a few bars, he threw his hands in the air and proclaimed: 'Madame, I play on the black keys, and then I try the white, but you sing in the cracks'.

This is something like the freedom that now exists in a certain kind of mathematics. It is known as fractal theory because it deals with dimensions but with a sublime and informed disregard of established conventions. Instead of the three dimensions, of Euclidian space (length, breadth and height, with all axes at right angles), we can use fractional dimensions, such as $3\frac{1}{2}$, or some other variant. To pass to the fourth dimension we have to get off halfway between three and four. This is what gives fractals their amazing versatility and interest.

Now, we have long been able to solve cubic equations, using Newton's method of iteration. That is to say, we set up a program to solve a cubic equation, say $Z^3 = 1$. We now understand that there are three roots; x = 1 (which is normally given as 'the answer'), along with two complex (imaginary) roots, $x = -.5 + .8666 \times \sqrt{-1}$, and $x = -.5 - .8666 \times \sqrt{-1}$. But we are not really interested in using the algebraic method of solving, only in producing a visual picture, or fractal, of the analysis and solution. (The fractal is also an imaginary number, which is why we need a visual, not a numerical, solution.) So, following Newton, we guess the answer as close as we can (Z_0), and substitute this value in the new equation $Z_1 = Z_0^3 - 1$. We repeat this procedure, continuing the iteration (say) 168 times. (This work is done by the computer.) Meanwhile, the various approximate solutions, and approaches to a solution, are printed out in different shapes and colours, each being shown as a spot of colour according to its value and the number of iterations to obtain it. When the program is over, the solutions, in sequence, are printed out as the most unusual shapes and combinations of colours. It is a new art form.

The interest of all this for cosmology at this time is that we are now talking about a new science, the science of 'chaos'. This started from the investigation of turbulence. Turbulence appears in liquids or gases affected by tremendous and continually changing random forces, for example in rivers or waterfalls 'boiling' as they strike violently against a jumble of rocks. Turbulence is also a feature of a liquid undergoing a change of state, for example, water in a kettle changing to steam by furious boiling at 100° Celcius. This is a universal principle.

Virtual and actual masses

In the past decade, the concept of virtual existence and virtual properties of matter (including mass) have made considerable headway in several areas of scientific thought. It reflects a powerful urge to give existential status to the Aristotelian construct of *potentia* in contrast to what is actually in being now, *actus*. For example, a personal computer is furnished with the physical object of a hard disk built in, as well as a separate 'floppy disc' drive to save materials and store them outside the computer on 'floppies'. These are used to set aside and preserve materials for special use later. We can also confuse or trick the computer by pretending we also have two extra *virtual* disc drives; we can even set aside a region of the memory for data and programs for these two imaginary discs. This 'trick' causes the computer to suffer constructive confusion. There is no way, except by printing them out on paper, that these materials can be taken out and physically examined (as 'floppy discs' or 'hard disks' can be). They exist in a self-preserving, quasi 'imaginary' mode until, given the correct signal, they manifest their presence by physically appearing. It is a little like resurrecting the souls of unbaptised infants from Limbo.

The false vacuum
This is the model I want to use in explaining the way the instant of creation works. We have to recognise the virtual dimension of reality where 'things' function *in potentia* and maybe *in actus* too, if they receive the right signal. The false vacuum is just such a thing. No false vacuum has ever been observed; it would in fact be impossible to do so. The change from the virtual to an actual vacuum (to give them a more meaningful title) would annihilate the virtual vacuum. The theory is that everything which came into being when the words 'Let there be Light' were spoken, already existed (perhaps in God's mind) but only in virtual form, *in potentia*. In physical reality, they existed, but only in what is called a 'false vacuum'. (It is false because it is virtually full of everything.)

The theory is predicated on certain ways of explaining subnuclear particles. A false vacuum, just like a real one, has no structure within, and no matter. What it does have is a large amount of positive energy, and a tremendous pressure – enough to blow the whole universe apart. It is like water changing to ice, like supercooled liquid nitrogen changing to a gas, or like doping a solution with crystals: a super-cooled or super-saturated solution is caused to crystallise. There is a phase change which, in this case, causes the universe to appear and suddenly (in fact instantly) expand almost to infinite size. Any particles or galaxies or star clusters which may appear abruptly, have been virtually present – and now become actual, following the Creator's command.

Our only comment here is to quote from S.A. Graves. He warned that: 'To deploy arguments at all *directly* against the manifestly absurd is to credit it with some intellectual dignity and to muffle its self-annihilating character.' So we will conclude here, leaving the matter an open question for time and the reader to decide.

Index

aboriginals, 7–9 *passim*
accelerators, 126, 128
Africa, 6, 7
air, 15–16, 21, 138
alpha rays, 124, 154
America, North, 6–7
ananke, 19
animism, 7, 187
Anselm, Saint, 191
anthropic principle, 188
'antimatter', xi, 10, 124, 154–6, 158, 165
Aquinas, Saint Thomas, 16, 20,
    189–92 *passim*
Arabs, ix, 9, 179, 189
Archimedes, 29, 107, 113, 114
Argand, Jean Robert, 204
Aristotle, 11, 15, 17–18, 20–4, 29–31
    passim, 87, 98, 166, 180, 189, 191,
    192, 205
armillary spheres, 12
astrology, 14
Atlantic, 18–19
atomism, 10, 114–17
atoms, x, 54, 79, 81–5, *82*, 113–32,
    140, 146
    'solar system', 83
    splitting, 124, 129–32, 179
    weights, 117–18, 121–3
Augustine, Saint, 145
Australia, 6–9 *passim*
Avicenna, 189, 190
Avogadro, Amadeo, 116, 118
Ayer, A.J., 74

Babbage, Charles, 58
Babylonians, ix, 9
Bacon, Francis, 20, 76, 171

baryons, 126
Becquerel, 154
Bell, Jocelyn, xi, 177–8
Berkeley, Bishop, 86, 88, 93, 180
beta rays, 154
Betelgeuse, 69
Beveridge, William, 194
'black body' experiment (Planck),
    80–1, 156
'black holes', 91, 119, 144–5, 203
Blake, William, 187
'blind beetle' analogy (Einstein), 104,
    138
Bohm, D., xi
Bohr, Niels, x, xi, 39, 75, 81–4, 86,
    87, 89, 90, 93, 134, 179–80
bomb, atomic, 3, 96, 123, 124, 129–30
Born, Max, x, 88
'bosons', 127, 156
Boyle, Robert, 115
    law, 183
Bridgman, Percy, 74
Broglie, Prince de, x
Brownian movement, 79, *80*
Bunsen, Robert Wilhelm, 49
Burbank, Luther, 165
Burnham, Robert, 64

calendar, 14–15
caloric, 10
Cannon, Annie, 65, 66
Carlyle, Thomas, 194
Carroll, Lewis, 85, 91
cathode rays, 80, 121–2, *122*
causality, xi, 39, 78, 83, 88, 92, 93,
    179, 180
Cavendish, Henry, 138, 139

Cayley, Arthur, 88
Chadwick, James, 131
Chang Yeu, 13
du Châtelet, Madame, 26
chemistry, 115–17
Chernobyl, 96
China, 4, 9–16, 21, 22, 113, 179
  astronomy, 12–14
Christianity, ix, x, 4, 19–23 *passim*,
  26, 30, 93, 186, 189–91; *see also*
  religion
chou pei Sundial theory, 12
classification, xi, 51
  stellar, 64–6, 73, 133, 135–6
Clerk Maxwell, James, 97, 100, 103
clocks, 98, 101–2
  caesium, 47
  pendulum, 33
  water, 31
Close, Frank, 113
Cocconi, Giuseppe, 177
Compton, Arthur Holly, 156
  'scattering', 157, 184
Confucianism, 22
Copernicus, 22–3, 26
  system, 22–4, 26
'cosmic hum', 145
Coulomb's law, 103, 176
Creation, ix, 38–9, 52, 134, 145–7,
  186–9, 195, 198–201 *passim*
  Big Bang theory, 121, 134, 137,
    146, 157, 200, 203
Crookes, Sir William, 94
Curie, Marie & Pierre, x, 53

Dalton, John, 79, 116–17
dark matter, 119, 134, 143–4, 160,
  168, 200, 202–3
Darwin, Charles, 165, 187–8, 192–4
Darwin, Erasmus, 58
Darwinism, 186
Davy, Sir Humphrey, 115
Democritus, 18, 21, 29, 113–15
  passim, 189
Descartes, René, 20, 34, 195
determinism, 77–8
Dirac, Paul, xi, 10, 88–91 *passim*, 124,
  154
Doppler, Johann Christian, 36–8
  effect, 36–7, *37*, 107
Drake, F.F., 174–7 *passim*

Draper, Henry, 66
dwarfs, 135, 136
dynamics, 30–8

Earth, 12, 13, 15, 21–6 *passim*, 35,
  39–42, 51–5, 108, 141–2, 174
eclipses, 11, 22, 30, 77, 78, 96, 134, 201
economy, 76, 166
Eddington, Air Arthur Stanley, 77,
  93, 201
Egyptians, xi, 113, 198
Einstein, Albert, x, 3, 33, 36, 39, 69,
  74–7 *passim*, 79, 81, 83, 84, 87–91,
  95–109, 124, 136, 137, 142, 154–6,
  179, 196–7, 201; *see also* relativity
electrolysis, 117
electromagnetism, 10, 97, 103, 140,
  155, 156
electrons, x, 39, 79, 81–4, 86, 88–90,
  117, 119–20, 123, 125, 126, 128–9,
  131, 134, 140, 154–8, 183–4
elements, 11, 15, 115–23, 138, 146
  atomic weight, 117–18, 121–3
  periodic system/table, 117–21, 146
  valency, 118
energy/mass equation (Einstein), 96,
  102, 154
Eotvos, Baron, 139
Epicurus, 114
Eratosthenes, 29, 114
ether, x, 16, 26, 67–9, 89, 97, 100–1,
  152
Euclid, 73, 102, 104, 105, 140, 198
European Centre for Nuclear
  Research (CERN), 128
evolution, 58, 165, 174, 186, 188, 192–4
exchange function, 164
exclusion principle (Pauli), x, 81, 126
expansion, of universe, 73–4, 107–8,
  133, 146–7

Faraday, Michael, 10, 103, 105–6, *106*,
  108, 115, 117, 142
Fechner's Law, 114
Fitzgerald, George Francis, 101
Fizeau, Armand Hippolyte Louis,
  66–7, *67*
Flagstaff Observatory, Arizona, 107,
  168
Flammarion, 168
Fleming, Williamina, 65–6

INDEX

Flexner, Abraham, 108
Fontenelle, Bernard Lebovier de, 163
forces, 10, 80, 103, 106, 130–1,
  137–40, 179, 199, 200
fossils, 52, 53, 199
fractal theory, 205
Franklin, Benjamin, 125
Frauenhofer, Joseph von, 49
  lines, 37, 49
French Revolution, 26, 51

galaxies, 58–61, 59, 73–4, 107–8,
  143–4, 146, 159, 174–7, 179, 200
Galileo Galilei, ix–x, 20, 22–4, 26,
  30–1, 47, 57, 73, 98, 114, 139, 179
  Dialogue, ix–x, 23–4, 30
gallium, 120
gamma rays, 102, 126, 154–8, 157
Gamow, George, 121, 146, 157, 198
gases, 115–16, 120, 183
  inert, 117, 120–1, 138
Gell-Mann, Murray, xi, 123
geology, 52–5
geometry, 14, 73, 104–5, 140
  space-time, 104–5
Georgi, H.M., 3
germanium, 120, 149
giants, 135, 136
Gilbert, William, 108
Gilson, Etienne, 185
gluon, 131
God, existence of, 186–97
gods/goddesses, 17, 19, 187
Graham's law, 183
Graves, S.A., 206
gravitons, 100, 127, 142, 158, 201
gravity, 33–6 passim, 47, 57, 96, 99,
  100, 103–9, 134, 136–9 passim, 137,
  140, 141–2
Greeks, ix, 4, 11–12, 17–26, 29–30,
  113–14, 179
  astronomy, 17
Green Bank formula, 174–6
Guth, Alan, 198, 202

hadrons, 126
Haldane, J.B.S., 167
Hawking, Stephen, 202
heat, 10, 80–1, 156
Hegel, Georg Wilhelm Friedrich, 10,
  180

Heisenberg, Werner, xi, 39, 86–90, 93,
  134, 179–80
helium, 41, 50, 89, 120–1, 138, 154
Henry's law, 183
Heraclitus, 24
Herschel, Caroline, 57
Herschel, John, 58
Herschel, William, 53, 57–8, 61, 133
Hertzsprung, Ejnar, 133, 135
  –Russell chart, 73, 133, 135–6,
  135
Hewish, Anthony, 153
Hoyle, Fred, 121, 146, 203
hsuan yeh (Brightness/Darkness
  theory), 12
Hubble, Edwin, 3, 73–4, 107–8, 146
  law, 107, 146
Hume, David, 85
hun thien theory, 13
Hutton, James, 53
Huxley, T.H., 10, 52
Huygens, Christiaan, 33–5, 34, 61, 79,
  103
hydrogen, 41, 83, 89, 117–19, 121,
  130, 138, 146

indeterminacy, 54, 83–4, 90, 92
Index of Prohibited Books, 23, 27
Inge, Dean W.R., 93
interferometer, 68–9, 68
  radio, 152–4
inverse squares, law of, 103, 176
ions, 117
isotopes, 121–3

Jansky, Karl, 149–52 passim
Jeans, Sir James, 93
Jesuits, 4, 9, 15, 27, 65
Jodrell Bank Observatory, 60, 151
John the Grammarian, 31
Joyce, James, 74, 123
Judaism, x, 27, 187, 189–90

Kant, Immanuel, 56, 85, 185, 195
Kapitza, Peter, 88, 124
Kelvin, Lord, x, 55
Kepler, Johannes, 31–2, 32, 35, 47
Kirchhoff, Gustav Robert, 49
'kiss of death', 158
Kropotkin, Prince, 193

LaMettrie, Julien Offray de, 34
Landau, Lev Davidovitch, 145, 198
Laplace, Pierre de, 52, 56, 77–8, 86, 87
Lavoisier, Antoine Laurent, 51
left-hand bias, 165
Leibnitz, Gottfried Wilhelm, 79
Lemaitre, Father, 147
lenses, gravity, 201–3, *201–2*
leptons, 126
life, 163–78
    definition of, 164–7
light, x, 11, 33, 36, 48–9, 66, 77, 79–81, 96, 100, 105, 136, 156, 184, 201–2
    'bending', 77, 96, 136, *137*, 201, *202*
    velocity of, 66–9, 99
Linde, Carl von, 202
Lippersley, Hans, 57
Lockyer, Sir Norman, 50
Lodge, Sir Oliver, 94, 148
logical positivists, 74
Lorenz, Hendrik Antoon, 101
Lovell, Sir Bernard, 151–2
Lowell, Percy, 167–9 *passim*, 173
Lucretius, 114

Mach, Ernst, x, 3, 29, 33, 73–7, 88, 89, 98, 101
Magellan, 140–1
magnetism, 98, 103, 105–6, *106*, 108–9, *109*, 140–2 *passim*
Maimonides, 189, 190
Manhattan Project, 123, 129
maps, 141
    star, 13, *13*
Marconi, Guglielmo, 148
Mariners, 171
Maritain, Jacques, 35
Mars, 41–3 *passim*, 56, 75, 168–73
Martineau, Rev. James, 185
Marx, Karl, 10
matrices, 88–90
matter, x, 113–32, 181–4
    dark, 119, 134, 143–4, 160, 168, 200, 202–3
    negative, 199
Maury, Antonia, 65, 66, 135
Mead, G.H., 73, 78
measurements, 31, 33, 35, 44–8
    'beacon' method, 60, 107

mechanics, Matrix, 89
    quantum, 89, 181
Mendeleyev, Dmitry Ivanovich, 117–18, 120–1
Mercator, Gerard, 141
Mercury, 42, 56, 97
    perihelion, 41, 97, *97*, 136
mesons, 126, 182
metaphysics, x, xi, 4, 17, 18, 29, 74, 90; *see also* religion
Michelson–Morley experiment, x, 67–9, *68*, 89, 97, 100, 101, 152
microscope, electron, 128–9
Michurin, Ivan Vladimirovich, 165
Milky Way, 12, 60, 73, 107, 127, 151, 158, 169, 174, 175
Minkowski, Hermann, 73, 102–3, 138
Mo Ching/Mohists, 10
Moon, 11, 30, 202
Morrison, Philip, 188
motion/movement, 10, 21–4, 29–38, 47–8, 55–6, 98–9, 165, 191–2; *see also* dynamics
    Brownian, 79, *80*
Mount Wilson Observatory, California, 107, 108, 133
myths, ix, 18–19, 38–9, 187, 196

Napoleon, 51–2
NASA, 75, 169, 172
natural selection, 165, 188, 193
nebulas, 12, 61–3, 107, 143, 159
    nebular hypothesis (Kant–Laplace), 56–7, 143
Needham, Joseph, 13
negative theology, 190
neon, 120–3 *passim*
von Neumann, John, 179
neutrons, 82, 84, 119, 120, 123, 125, 127, 128, 130, 131, 140
Newton, Sir Isaac, x, 10, 11, 33, 35–6, 47–9, 73, 78–80, 95, 96, 98, 105, 114–15, 136, 139, 166, 179, 180, 190, 201, 205
novae, 144
nuclear fission, xi, 102, 129
    force, 140
    fusion, 130–1, 146
    power, 129–32
numbers, imaginary, 203–4, *204*
    random, 87–8, *87*

INDEX

Occam's razor, 76
Oersted, Hans Christian, 103
Omar Khayyam, 204
ontological argument, 191
operationism/Machism, 74–7, 89
Oppenheimer, J. Robert, 95
Ozma Project, 176

Paley, William, 192–4 *passim*
Palomar Observatory, 133
Paracelsus, 115
parallax, 46–7, *46*
Parmenides, 180
particles, 11, 79, 80, 102, 114, 124–9, 135, 142, 179–84
    subatomic, x, xi, 3, 84, 90, 91, 101, 102, 123, 124, 154, 181–3
    passim, 182
'pathetic fallacy', 170, 177
Pauli, Wolfgang, x, 81, 126
Penrose, Roger, 144
Penzias, Arno, 146, 157, 202
phlogiston, 10
Phoenicians, ix
photoelectric effect, 79, 81, 124, 156, 184
photons, 79, 81, 127, 142, 144, 154, 156–60, 182
Pickering, Edward C., 65–6
Planck, Max, 3, 73, 77, 80–1, 137, 156, 179
planets, ix, 12, 14, 25–6, 30–3, 40–4, 46, 48, 55–7, 168–9, 174; *see also individual headings*
Plato/Platonism, 4, 11, 17–21, 23, 30, 77, 86, 87, 92, 189
Pluto, 32, 40, 41, 44, 168
Poincaré, Henri, 73, 97
porphyrin, 166
positrons, 154, 158
prehistory, 5–9
Priestley, J.B., 100
    Joseph, 115
probability, law/theory of, x, 85, 87, 180, 193
protons, 82–4 *passim*, 117, 119, 120, 123–8 *passim*, 131, 140, 154, 156, 182
    anti-, 154
Ptolemy, 13, 15, 21–6 *passim, 25*
    Almagest, 24–5

pulsars, 62, 153, 159, 178
Pythagoras, 23, 189

quadrant, 57
quantum theory, 74, 77, 80–1, 83, 84, 88–91, 93, 156, 180–4
quarks, 113, 123–9, 134, 140
quasars, 159, 177

radar, 47, 137, 151
radiation, 81, 122, 146, 147, 154–60, 202
radio, 60, 148–54, 156, 176–8
    astronomy, 151–2
    Cambridge Catalogue, 153
radioactivity, 53–5, 99, 140
radium, x, 53–4, 86, 98, 154
Ramsay, Sir William, 50, 120
Reber, Grote, 151, 152
relativity, x, 69, 73, 77, 90, 91, 93, 95–109, 134, 136–7
religion, 17, 27, 35, 52, 93, 185–97; *see also* Christianity
reproduction, 164–5
Ricci, Father Matteo, 15–16, 21–3 *passim*
Riemann, Bernhard, 73, 88, 97, 104
Roentgen, Wilhelm, 154, 158
Romans, 27
Roosevelt, F.D., 194
Rubin, Vera, 203
Rushdie, Salman, 189
Russell, Bertrand, 179
Russell, Henry Norris, 133, 135–6
Rutherford, Ernest, x, 81, 83, 86, 117, 124, 131, 154
Ryle, Martin, 153

Sagan, Carl, 167
Sakharov, Andrei, 155–6
satellites, space, 137, 158, 159, 169, 171
scandium, 120
Schiaparelli, Giovanni Virginio, 167–8
Search for Extra Terrestrial Intelligence (SETI), 178
Secchi, Father, 65
Shkhlovsky, Josif Samuilovich, 169
sighting tubes, 12
'singularity', 121, 145–7, 195, 199
Slipher, Vesto, 3, 73–4, 107, 133, 146
Smith, William, 45

Socrates, 17, 18
solar system, 23, *25*, 26, 30, 31, 40–4,
    55–8, 129, 137
Soviet Union, 117–18, 120–1, 137
space travel, 3, 75, 87, 137, 171–3
spectra, x, 37–8, *38*, 49–50, 64–6, 83,
    135
spectroscope, 37, 49, 64, 75
Spencer, Herbert, 188
Spinoza, Benedictus de, 95, 189, 190
spiritualism, 75, 94
Sputnik, 137
stars, 12, 14, 36, *37*, 40, 61–6, 107,
    133, 135–6
    classification, 64–6, 73, 133, 135–6;
        Harvard Project, 65–6
    clusters, 61–3 *passim*, 135
    spectra, 37–8, *38*, 64–6, 135
static, 108, 125, 149
'steady state' theory, 146, 203
Stoics, 24
string models, 198–201
Strutt, William (Lord Rayleigh), 120
Sun, 11, 21–3 *passim*, 26, 40, 41, 56,
    91, 96, 121, 129, 150, 158
sunspots, 20, 26, 30, 150
supernovae, 61, 62, 134, 144
symmetry, 200

teleological argument, 192
telephone, 149–50
telescope, 3, 12, 26, 30, 57–8, 75, 107,
    133–4, 163, 201
    radio, 152–4
Tertullian, 147, 195
Thales of Miletus, 78, 113
theodolite, 45
thermodynamics, second law of, x,
    106, 166–7, 182
Thomson, J.J., x, 81, 117, 121
'thought experiments', 75–6, 84–6,
    101, 102
time dilation, 101
Tipler, Frank, 178
Titius–Bode law, 56

topology, 104–5, 109
torsion balance, 139, *139*
totemic view, 5–9 *passim*
Trapezium, 61
triangulation, 45–6, *45*
turbulence, 86, 205
'twin paradox', 101–2

ultra-violet rays, 156
uncertainty principle (Heisenberg), xi,
    39, 85–9, 93
unified field theory (Einstein), 134,
    137–8, 140
'uniformitarianism', 53
United States, 75, 137, 169, 172
uranium, 54, 86, 122, 123, 129–30
Uranus, 32, 43–4, 58
Ussher, Archbishop, 52

vacuum, false, 206
Vasquez, Miguel, 56
Venus, 30, 41, 42, 56, 137
da Vinci, Leonardo, 31
Vienna Circle, 74
Voltaire, 26, 163

Watson-Watt, Sir Robert, 151
waves, x, 10, 11, 34, 36–7, 79–81, 100,
    105, 142
    gravity, 105–7, 201
    -particles, 80–1
    radio, 60, 144, 149–56, 176
weightlessness, 106, 141
'white holes', 144
Wilson, Robert W., 146, 157, 202
Wimshurst machine, 125
Wollaston, William Hyde, 49

X-rays, 154–6, 158–60, *159*

Yin and Yang, 10
Young, Thomas, x, 184

Zeno, 180, 185
Ziman, John, 51